D1014796

The Roles of Representation in School Mathematics

The Roles of Representation in School Mathematics

2001 YEARBOOK

Albert A. Cuoco
2001 Yearbook Editor
Education Development Center, Inc.
Newton, Massachusetts

Frances R. Curcio
General Yearbook Editor
Queens College of the
City University of New York
Flushing, New York

National Council of Teachers of Mathematics
Reston, Virginia

1906 Association Drive, Reston, VA 20191-9988
(703) 620-9840; (800) 235-7566; www.nctm.org

Library of Congress Cataloging-in-Publication Data:

Cuoco, Albert A.
The roles of representation in school mathematics / Albert A. Cuoco, Frances R. Curcio.
p. cm. — (Yearbook ; 2001)
Includes bibliographical references.
ISBN 0-87353-495-6
1. Mathematics—Study and teaching. 2. Mental representation. I. Curcio, Frances R. II. Title. III. Yearbook (National Council of Teachers of Mathematics) ; 2001.

QA1 .N3 2001
[QA11]
510′.71—dc21

2001030155

The publications of the National Council of Teachers of Mathematics present a variety of viewpoints. The views expressed or implied in this publication, unless otherwise noted, should not be interpreted as official positions of the Council.

Printed in the United States of America

Contents

Part 4: The Role of Context

Preface

> I had a scheme, which I still use today when someone is explaining something that I'm trying to understand: I keep making up examples. For instance, the mathematicians would come in with a terrific theorem, and they're all excited. As they are telling me the conditions of the theorem, I construct something which fits all the conditions. You know, you have a set (one ball)—disjoint (two balls). Then the balls turn colors, grow hairs, or whatever, in my head as they put more conditions on. Finally, they state the theorem, which is some dumb thing about the ball which isn't true for my hairy green ball, so I say, "False!"
>
> —Richard Feynman,
> *Surely You're Joking, Mr. Feynman!*

When mathematics teachers at any level get together to talk about what they do, two questions are almost sure to come up:

- What do we teach?
- How do we teach?

Questions about content and pedagogy are central to what we do. It is right that these two questions are so important; thinking about them leads to improved curricula and teaching methods.

But there's a third important question, one that occupies the careers of many educational theorists, that is beginning to make its way into discussions in teachers' lounges and department meetings:

- How do students learn?

In many ways this is a much more difficult question. It requires that we look into the minds of our students and that we think about things from *their* perspectives. It is very hard for an adult, experienced in mathematics, to assume the perspectives of a beginner. But many teachers, mathematicians, and educators are realizing that smart decisions about content and pedagogy require that we understand much more about the ways our students learn mathematics, how they come to develop mathematical habits of mind, and even how they develop misunderstandings about our discipline.

This yearbook is about one aspect of how students learn mathematics. More precisely, this book is about how students learn to build mathematical representations of phenomena. This year marks the twenty-fifth anniversary of this series of yearbooks, and it is appropriate that we take up this timely theme.

All of us have an intuitive idea of what it means to represent a situation; we do it all the time when we teach or do mathematics. We represent numbers by points on a line or by rows of blocks. We use equations and geometric figures to represent each other. We talk about numerical, visual, tabular, and algebraic representations. And we *think* about things using "private" representations and mental images that are often difficult to describe.

But what do we mean, precisely, by "representation," and what does it mean to represent something? These turn out to be hard philosophical questions that get at the very nature of mathematical thinking.

I believe that as mathematics itself evolves, new methods and results shed light on such questions—that mathematics is its own mirror on the very thinking that creates it. And sure enough, there is a *mathematical* discipline called representation theory. In representation theory, one attempts to understand a mathematical structure by setting up a structure-preserving map (or correspondence) between it and a better-understood structure. There are two features of this mathematical use of the word *representation* that mirror uses of "representation" in this book:

- The representation is the *map*. It is neither the *source* of the representation (the thing being represented) nor its *target* (the better-understood object). When a child sets up a correspondence between numbers and points on a line, the points are not the representation; the representation lives in the setting up of the correspondence.
- Representations don't just match things; they preserve *structure*. Entering on a calculator an algebraic expression that stands for a physical interaction is not, all by itself, a representation. If algebraic operations on the expression correspond to transformations of the physical situation, *then* we have a genuine representation. Representations are "packages" that assign objects and their transformations to other objects and *their* transformations.

The articles in this book present a wide array of perspectives about the nature of representations, how students create them, and how they learn to use them. The book is divided into four parts, each a collection of articles that deal with a related circle of ideas.

The first part, "Roles for Representations," sets the stage by providing a discussion of two central dialectics in the educational theory of representations:

- *Internal and external representations.* External representations are the representations we can easily communicate to other people; they are the marks on the paper, the drawings, the geometry sketches, and the equations. Internal representations are the images we create in our minds for mathematical objects and processes—these are much harder to describe.

Gerald Goldin and Nina Shteingold discuss this distinction in their opening article. They present an overview of the theoretical issues and discuss an approach that integrates the research on internal and external representations.

- *"Invented" and "presented" representations.* Fran Curcio first used these words to describe these types of representations; she means the difference between representations that students invent and those passed down from teachers. We hear a great deal these days about student-invented representations. These are often quite different from the classical representations that have evolved in mathematics. The article by Constance Kamii, Lynn Kirkland, and Barbara Lewis makes a strong argument for the importance of allowing students to develop their own representations. However, centuries of evolution have produced standard mathematical representations that have been used to solve extremely deep problems. Mark Saul struggles in his article with the challenges of helping high school students understand the "standard" representational systems—the symbols and operations —of algebra. And the article by Rina Zazkis and Karen Gadowsky looks at the difficulties undergraduates have exploiting the "hidden meaning" in representations built up from the ordinary representations of arithmetic.

The second part, "Tools for Thinking," discusses representations as devices people use to help them gain insights into mathematical phenomena. Irene Miura describes a fascinating connection between one's natural language and how one thinks about numbers and numeration. Michelle Stephan, Paul Cobb, Keono Gravemeijer, and Beth Estes give an approach to the "invent or present" tension that introduces standard measuring tools in response to students' needs. Carmel Diezmann and Lyn English look at the role of diagrams in doing mathematics and discuss strategies for helping students become proficient at inventing and using diagrams. Marty Schnepp and Ricardo Nemirovsky describe the tools they use to represent some subtle ideas treated in AP calculus; one byproduct of their approach is that students see the computational techniques of calculus as tools for solving problems about rate and accumulations. Mark and Maxine Bridger develop an alternative to the rule of three for describing real-valued functions of a real variable; their "mapping diagram" representation highlights some important features of functions that are often hidden by tabular, symbolic, and graphical representations. Daniel Scher and Paul Goldenberg take us on a dynamic tour of the law of cosines, and they show how interactive geometry environments can be used to represent, illustrate, and even discover this important theorem. And Larry Lesser catalogs some beautiful representations that help make sense of a counterintuitive situation in statistics called Simpson's paradox.

Part 3 is called "Symbols and Symbol Systems." Mathematics is full of symbols—symbols that stand for numbers, functions, geometric objects, even other symbols. But the symbols of mathematics aren't just aliases. They are part of symbol *systems* that allow people to act on and transform the symbols in meaningful ways. Susan Lamon describes her research into effective ways for children to represent, use, and calculate with rational numbers. A great deal of research has gone on around the use of algebraic symbol systems to represent and transform algebraic functions. Wendy Coulombe and Sarah Berenson describe their work with beginning algebra students in this area. Alex Friedlander and Michal Tabach describe their work around multiple representations, using algebraic symbolism as one of several mechanisms for describing functions. Deborah Franzblau and Lisa Warner investigate different symbol systems for describing recursively defined phenomena, contrasting, for example, subscript and functional notation. Finally, Regina Kiczek, Carolyn Maher, and Robert Speiser tell the story of a student with whom they worked over the course of several years and who developed some creative ways for using the binary system of enumeration to solve a combinatorial problem.

The last part, "The Role of Context," looks at the interplay between modeling and representation. Kristine Reed Woleck offers an insightful look at her young students' work, showing how their representations of mathematical situations as pictures evolve into symbolic representations. Phyllis and David Whitin describe their work using literature with children to elicit mathematical thinking. Margaret Meyer picks up the story at the middle school level and shows how pictures and icons can be incorporated into symbol systems that closely approximate the classical system of algebra. Michal Yerushalmy and Beba Shternberg describe their approaches to strengthening what they call the "fragile link" between the visualization of situations and the concept of function and to helping students develop skill at using classical algebraic symbolism. And Josh Abrams takes us inside his high school modeling course, describing the techniques he uses to teach explicitly the skills of modeling and representation.

Assembling this collection of articles and helping authors revise their drafts were the work of an expert Editorial Panel:

- Hyman Bass, University of Michigan
- Carolyn Kieran, Université du Québec à Montréal
- Arthur B. Powell, Rutgers, The State University of New Jersey—Newark
- Jesse Solomon, City on a Hill Public Charter School, Boston, Massachusetts

Frances Curcio, the general editor for the 1999–2001 yearbooks, helped us through the entire process. Fran participated in the editorial deliberations, helped us stay on task, furnished us with context and background, and dealt with every detail at every level, all at once, all the time. Fran, the four panelists, and I quickly formed a team; we built on one another's ideas, learned from one another, and collaborated in ways that I'll very much miss.

Several other people worked behind the scenes to make this book possible. Helen Lebowitz and Sara Kennedy worked with me to communicate with authors and to manage the substantial amount of correspondence involved in producing this book. They provided exactly the help I needed with everything from scheduling meetings to editing articles. Wayne Harvey offered support, advice, and expert editing suggestions. Charles Clements and the NCTM staff worked incredibly hard on this project, editing and advising along the way and contributing to every aspect of the book. And my wife, Micky, helped me manage the details (a task at which I'm notoriously inept), listened to several hundred variations of my saying "It's almost done," and (almost) never complained about the meetings and the late nights and the piles of manuscripts on the floor of our study.

This book would not have been possible without the contributions of everyone who submitted a manuscript. The hardest part of this job was selecting the final manuscripts; given more pages, I would have liked to include much more than what is here. As I read the drafts, I was struck at how much knowledge, insight, creativity, and common sense are distributed across our field. And all this expertise—mathematics, pedagogy, and epistemology—gets integrated, synthesized, and applied every day in thousands of classrooms all over the country by classroom teachers—teachers who know how to take the ideas in this book and turn them into classroom experiences that make young people see the beauty and excitement in mathematics.

Albert A. Cuoco
2001 Yearbook Editor

1

Systems of Representations and the Development of Mathematical Concepts

Gerald Goldin

Nina Shteingold

IDEAS about representation in the teaching and learning of mathematics have evolved considerably in recent years, with contributions by both researchers and practitioners. In this article, we offer an informal introduction to some of the fundamental concepts. They pertain broadly to mathematics, the psychology of mathematical learning and problem solving, children's mathematical growth and development, the classroom teaching of mathematics, and the rapidly changing technological environment in which mathematics education is taking place (Goldin and Janvier 1998a, 1998b; Goldin and Kaput 1996; Janvier 1987). We shall also contrast these fundamental ideas with other perspectives, especially those of behaviorism and constructivism, in framing the important issues.

One of our aims is to understand better the blocks, or cognitive obstacles, that students may have to acquiring certain mathematical concepts. We consider examples drawn from selected task-based interviews with young children, illustrating some of their early understandings of negative numbers. Here specific obstacles may arise from, or be associated with, the particular representations that students use. We describe some interesting and contrasting obstacles of this sort and explore how such impasses are overcome as more efficient, powerful, or streamlined representations develop.

Let us remark immediately that a mathematical representation cannot be understood in isolation. A specific formula or equation, a concrete arrangement of base-ten blocks, or a particular graph in Cartesian coordinates makes sense only as part of a wider *system* within which meanings and conventions have been established. The representational systems important to mathematics and its learning have *structure*, so that different representations within a system are richly related to one another.

It is also important to distinguish *external* systems of representation from the *internal*, psychological representational systems of individuals. Both need our attention. External systems range from the conventional symbol systems of mathematics (such as base-ten numeration, formal algebraic notation, the real number line, or Cartesian coordinate representation) to structured learning environments (for example, those involving concrete manipulative materials or computer-based microworlds). Internal systems, in contrast, include students' personal symbolization constructs and assignments of meaning to mathematical notations, as well as their natural language, their visual imagery and spatial representation, their problem-solving strategies and heuristics, and (very important) their affect in relation to mathematics.

The *interaction* between internal and external representation is fundamental to effective teaching and learning. Whatever meanings and interpretations the teacher may bring to an external representation, it is the nature of the student's developing internal representation that must remain of primary interest. We highlight this interaction with some examples of young children's acts of assigning mathematical meaning as they consider structured, external task representations and construct their personal, internal representations of signed numbers.

Along the way, we note some important connections that may occur among distinct representations or systems of representation. These include uses of analogy, imagery, and metaphor (English 1997) as well as structural similarities and differences across representational systems. Although mathematics is an exceptionally precise discipline, there is inevitably some *ambiguity* present in both external and internal representations, and this ambiguity may also be a source of cognitive obstacles. Frequently it is the contextual information that permits the resolution of such ambiguity when it occurs.

An important motivation for our work on representation is the goal of greatly increasing the proportion of schoolchildren who succeed at a high mathematical level. Our perspective is that the vast majority of students are not inherently limited in their ability to understand mathematical ideas—including advanced concepts of algebra and geometry. This point of view has a long and noble history, with such advocates as Maria Montessori (1972, 1997), Jerome Bruner (1960, 1964), Zoltan Dienes (1964, 1972), and Robert Davis (1966, 1984). We think it is not a coincidence that these educators shared a focus on introducing mathematical ideas to children through the exploration of carefully structured task representations.

The universality of access to mathematical achievement is also explicit in the goals expressed by the National Council of Teachers of Mathematics

(2000) and in many current state curriculum standards. But relatively few schools have found and implemented the techniques that will achieve breakthroughs for large numbers of children. Many people, from parents to research mathematicians, remain skeptical that such visionary goals are possible and continue to see all but the most routine mathematics as the exclusive province of students with innate talents. Without denying the existence or importance of ability differences in mathematics, we suggest here that the apparent limitations in some children's understandings are *not intrinsic.* Rather, they are a result of internal systems of representation that are only partially developed, leaving long-term cognitive obstacles and associated affective obstacles.

Impasses of this sort persist as long as the representational tools for overcoming them are absent. But the tools can be acquired when we focus explicitly on them in the learning process. This leads us to consider that the fundamental goals of mathematics education include representational goals: the development of efficient (internal) systems of representation in students that correspond coherently to, and interact well with, the (external) conventionally established systems of mathematics.

EXTERNAL SYSTEMS OF REPRESENTATION

A representation is typically a sign or a configuration of signs, characters, or objects. The important thing is that it can stand for (symbolize, depict, encode, or represent) something other than itself. For example, the numeral 5 can represent a particular set containing five objects, determined by counting; or it can stand for something much more abstract—an equivalence class of such sets. It can also represent a location or the outcome of a measurement. A Cartesian graph is likewise a representation. It can depict a set of data, for example, or it can represent a function or the solution set of an algebraic equation. So we see that the thing represented can vary according to the context or the use of the representation. The numeral and the Cartesian graph are examples of what we are calling *external representations* in mathematics—students can produce them, and we can point to them in classrooms and discuss their meanings.

Such representations do not stand alone. The numeral 5 belongs to the conventional symbol system for arithmetic that begins with the individual signs 0, 1, 2, 3, and so on, and includes the multidigit base-ten numerals, the symbols for arithmetic operations and equality, the conventions for representing fractions, and so forth. The orthogonal pair of real-number lines and the identification of a point in the plane with a pair of real numbers provides a system of rules for creating Cartesian graphs that is extremely useful and flexible. Indeed, the individual numeral, or the particular graph, is almost meaningless apart from the system to which it belongs. Systems of external representation are *structured* by the conventions that underlie them. Once a system is established, patterns in it are no longer arbitrary; they are "there to be discovered," though greater flexibility of interpretation is present than is commonly realized.

One important aspect of representation is its two-way nature. The representing relation—depiction, encoding, or symbolization—often can go in either direction. Thus, depending on the context, a graph (e.g., a circle of radius 1 centered at the origin in the Cartesian plane) could provide a geometrical representation of an equation in two variables (e.g., the equation $x^2 + y^2 = 1$). Alternatively, an equation relating x and y could provide an algebraic symbolization of a Cartesian graph.

Much of the history of mathematics is about creating and refining representational systems, and much of the teaching of mathematics is about students learning to work with them and solve problems with them (Lesh, Landau, and Hamilton 1983). Some external systems of representation are mainly notational and formal. These include our system of numeration; our ways of writing and manipulating algebraic expressions and equations; our conventions for denoting functions, derivatives, and integrals in calculus; and computer languages such as Logo. Other external systems are designed to exhibit relationships visually or spatially, such as number lines, graphs based on Cartesian, polar, or other coordinate systems, box plots of data, geometric diagrams, and computer-generated images of fractals. Words and sentences, written or spoken, are also external representations. They can denote and describe material objects, physical properties, actions and relations, or things that are far more abstract.

The traditional representational systems of mathematics that we have mentioned are *static* in the sense that they provide rules or frameworks for creating *fixed* external formulas, equations, graphs, or diagrams. But new technologies, from graphing calculators to computer-based microworlds, present a world of new, *dynamic* possibilities—external systems where representations can be changed dramatically with the click of a mouse or the drag of a cursor and linked automatically and continuously to one another (Kaput 1989). Thus the student can watch a graph change as a parameter in a symbolic expression represented by the graph is adjusted. The power of this technology for teaching is only beginning to be tapped.

INTERNAL, PSYCHOLOGICAL REPRESENTATION BY STUDENTS

External representational systems are useful, or limited in usefulness, according to how individuals understand them. For instance, some students manipulate mathematical expressions well, skillfully performing arithmetic and algebraic computations. But even a high level of ability to do this need not imply an understanding of mathematical meanings, the recognition of structures, or the ability to interpret the results. Rules in mathematics can be learned and followed mechanically—and definitions can be memorized—without very much conceptual development having taken place.

How, then, can we describe students' understandings of a mathematical concept? Is the configuration –3, for instance, understood and interpreted in the desired variety of ways? Perhaps it is seen merely as a mark or a minus sign, followed by the numeral 3. One student may have formed some meaningful, related concepts but failed to associate them with the symbolic notation. Another student may have little or no notion of negative numbers or even see numbers less than zero as being impossible. To characterize the complex cognitions that can occur, one needs a model or a framework. One approach is to consider and try to describe the internal representations or, as they are sometimes called, "mental representations" of the student (Kosslyn 1980; Palmer 1977).

Systems of internal cognitive representation can be of several different kinds (Goldin 1987, 1998a). *Verbal/syntactic* representational systems describe individuals' natural language capabilities—mathematical as well as nonmathematical vocabulary and the use of grammar and syntax. *Imagistic* systems of representation include visual and spatial cognitive configurations, or "mental images." These contribute greatly to mathematical understanding and insight. Imagistic systems also include kinesthetic encoding, related to actual or imagined hand gestures and body movements, which are often important to capturing the "feel" of the mathematics. Likewise auditory and rhythmic internal constructs are essential, as children learn letters and counting sequences, clap hands in rhythm, and so forth.

Formal notational representation also takes place internally, as students mentally manipulate numerals, perform arithmetic operations, or visualize the symbolic steps in solving an algebraic equation. *Strategic and heuristic* processes for solving mathematical problems are represented as children develop and mentally organize methods such as "trial and error," "establishing subgoals," or "working backward." These representations, although highly structured, can sometimes be quite unconscious—the child making effective use of a strategy may find it difficult to explain how he or she is approaching the problem.

In addition, intertwined deeply with cognition, we have individuals' *affective* systems of representation. These include students' changing emotions, attitudes, beliefs, and values about mathematics or about themselves in relation to mathematics. Affect can importantly enhance or impede mathematical understanding (DeBellis 1996; Goldin 2000; McLeod and Adams 1989).

We cannot, of course, observe anyone's internal representations directly. Rather, we make *inferences* about students' internal representations on the basis of their interaction with, discourse about, or production of external representations. Skilled teachers do this almost automatically, paying attention to their students' words, written work, use of manipulative materials, or use of calculators and computers as they try to understand individual conceptions and misconceptions. It is sometimes useful to think of the external as representing the internal, as when a student draws a diagram or writes a formula to describe what he or she is thinking. Simultaneously we can think of the internal as representing the external, as when a student formulates a "mental picture" of operations described in an arithmetic formula. This shifting perspective is characteristic of the two-way nature of representation.

Internal representations do not simply encode or represent what is external. In using them to characterize individuals' conceptual understandings, we stress that internal representations can refer to *each other* in complex ways. This is one of their most important psychological aspects. Thus an individual's internal representation of –3 may include verbal phrases, such as "the opposite of three" or "the negative of three"; complicated visual or spatial images, such as a location three spaces to the left of zero on a number line; kinesthetically encoded images or action sequences, like taking three steps backward or turning around and taking three steps; various related formal notational procedures, such as combining –3 with +3 to obtain 0; and many other possibilities. Such configurations can, at various times, represent each other, with words representing a visual image, an image representing a formal procedure, and so on.

A mathematical concept is learned and can be applied to the extent that a variety of appropriate internal representations have been developed, together with functioning relationships among them. We infer the nature of the developing representations, and their adequacy, in part from the individual's interactions with the external, conventionally developed systems of representation of mathematics and in part from his or her interactions with nonmathematical situations.

Different researchers have focused on different aspects of the role of representation in mathematics learning and teaching. Some have emphasized representation in particular mathematical domains, such as integers (Carraher 1990), functions (Even 1998), or geometry (Mesquita 1998). Others have studied the processes whereby children working in groups construct representations (Davis and Maher 1997). A major, developing field of inquiry is the role of pictorial imagery, analogy, and metaphor in internal representation (English 1997; Presmeg 1992, 1998), a study very much in the spirit of continuing George Polya's ideas about mathematical problem solving (Polya 1954a, 1954b). Still others have done work on embodying mathematical ideas through dynamic, linked computer representations (Edwards 1998; Kaput 1989, 1992, 1993). In this paper we have sought to provide a synthesis of these views and to include additional references that can assist the interested reader (Goldin 1998b).

A Comparison of the Representational Perspective with Behaviorist and Constructivist Perspectives

The representational perspective taken here may be contrasted with other points of view. Two broad perspectives, quite different from each other, that spurred research and greatly influenced classroom practice in the past few decades came to be known as *behaviorism* and *constructivism.*

The psychological school of behaviorism, ascendant during the 1950s and 1960s, aimed to explain learning entirely through external, observable variables—structured stimulus situations, responses, and the reinforcement of desired behaviors (Skinner 1953, 1974). Behaviorists sought to avoid inferences about internal cognitive states. In school mathematics, behaviorists tended to focus on students' acquisition of skills, rules, and algorithms, with more complex systems of responses built up out of these. Classroom goals were framed as "behavioral objectives" or "performance objectives," which are very compatible with the measurement of achievement through standardized tests (Sund and Picard 1972).

Behaviorism was challenged by ideas from developmental psychology and cognitive science, within which we highlight the constructivist school that gained adherents during the 1980s. Here, all knowledge was seen as constructed from the individual's subjective world of experience (von Glasersfeld 1991). Constructivists thus placed great emphasis on the internal (as opposed to the behaviorists' focus on the external). They tended to reject the idea of "objectivity," even in mathematics, and relied on the coherence of discourse among individuals for the viability of their interpretations. Social constructivists placed special emphasis on the cultural and sociological processes through which knowledge is formulated (Ernest 1991).

In a sense, these two distinct philosophies connect with the different philosophies of teaching and public education, whose clash has evoked some controversy in recent years. On the one hand are those who favor basic mathematical skills, correct answers through correct reasoning, individual drill and practice, more direct models of instruction, and measures of achievement through objective tests. They tend to prefer the behaviorist characterization of skills and see constructivism as far too subjectivist. On the other hand are those who value children's making their own discoveries in mathematics, open-ended questions that may have more than one answer, different conceptualizations and interpretations by different children, less use of teacher-centered models of instruction, group as well as individual problem-solving activity, and alternative assessments. They tend to regard constructivism as the preferred research base, seeing the behaviorist approach as far too objectivist.

The research on representation, as we have described it, bridges the gap. It involves explicit focus on both the external and the internal, with the utmost attention given to the interplay between them. Through interaction with structured external representations in the learning environment, students' internal representational systems develop. The students can then generate new external representations. Conceptual understanding consists in the power and flexibility of the internal representations, including the richness of the relationships among different kinds of representation. Thus the research on representation achieves a synthesis of the other two perspectives, drawing on the best insights each offers without dismissing the contributions of the other.

This lends itself to a more inclusive educational philosophy—one that values skills and correct answers as well as complex problem solving and mathematical discovery, *without seeing these as contradictory*. The goal is high achievement in mathematics, for the vast majority of students, through a variety of different representational approaches.

RELATIONSHIPS AMONG REPRESENTATIONS

Next we comment on the important psychological roles of analogy (or simile) and metaphor. These come into play when images are evoked to develop, explain, or interpret mathematical constructs. Such processes always involve more than one system of representation—as when one type of construct is asserted as "being like," or as "being," another construct.

Returning to the example of negative numbers, a child may think that –3 is "like giving away three dollars" or "like owing three dollars," without quite specifying how or why it is like that. One representation, the symbolic notation –3, is interpreted here in relation to another, the real-life imagery associated with giving away or owing money. In a sense, to think of –3 as "being a number" extends metaphorically the meaning of "number." Negative quantities do not fit an earlier-developed cardinal meaning of "number" as the outcome of counting the elements in a finite set. The label "number" works here because many things that can be done with "numbers" can also be done with "negative numbers." Nevertheless, mathematicians into the eighteenth century were loath to accept negative numbers as "existing"; they were considered to be merely a shorthand for ordinary numbers oriented in an opposite direction.

Even to say that "zero is a number" or "a fraction is a number" can have such a metaphorical aspect. Early mathematicians were suspicious of both assertions. Indeed, the history of mathematics is replete with examples of new constructs regarded at first as unreasonable, bizarre, or not really existing—and names like "irrational" and "imaginary" were given to such numbers. But as new and better representations were constructed for these systems (e.g., the representation of imaginary numbers in the complex plane or by 2×2 matrices, or the association of irrational real numbers with Dedekind cuts of the rationals), the world of mathematics became more comfortable with them.

The extension of meanings in these ways is often associated with representational systems eventually embedded in larger ones. Positive and negative numbers, rational and irrational numbers, are all represented by the construct of the real number line, which is in turn embedded in the complex plane. The number line as an external system of representation allows us to highlight easily the *ordinal* relationships among numbers. Lesser numbers are to the left, greater numbers to the right, and the notion of one number being "between" two others has an immediate spatial interpretation. The *cardinal* notion of number as descriptive of a class of sets is no longer fundamental.

But negative whole numbers can also be represented cardinally. In such a representation, based on "signed cardinality," the expression $-n$ means a set of n objects or counts tagged as being "less than zero." A familiar classroom representation based on signed cardinality is to allow black chips to stand for positive units and red chips to stand for negatives. Reds and blacks can be created, or introduced in pairs; or they can annihilate each other in pairs. This (semiconcrete) system can also be used to represent operations with signed numbers. Adding positive numbers is to put in black chips, adding negative numbers puts in red chips, and so forth.

Effective mathematical thinking involves understanding the relationships among different representations of "the same" concept as well as the structural similarities (and differences) among representational systems. That is, the student must develop adequate internal representations for interacting with various systems. This entails assigning appropriate meanings, performing appropriate mental operations in interaction with external, structured systems, and being able to resolve the inevitable ambiguities when they occur.

Thus we conjecture that a fully developed concept of signed number involves internal representations both for number line and signed cardinality interpretations, whereas partially developed concepts may function in one system but not another. We provide an indirect confirmation of this conjecture in the observations described below of situations where children are able to manipulate signed numbers meaningfully in one representational context and not in another.

COGNITIVE OBSTACLES

The notion of cognitive representation helps us understand some of the blocks or obstacles that children may have to particular mathematical ideas.

How do such blocks result from, or connect with, the particular representations the children are using? How can cognitive obstacles be overcome through the increasing power of a representational system or through new systems brought to bear on concepts that are already partially developed?

New, internal systems of representation are typically built up from preexisting systems. It is helpful to identify three main stages in their development (Goldin and Kaput 1996):

1. First is an inventive-semiotic stage (Piaget 1969), during which new characters or symbols are introduced. They are used to symbolize aspects of a previously developed representational system, which is the basis for their "meaning."
2. In the second stage the earlier system is used as a kind of template for the structure of the new system. Rules for the new symbol-configurations are worked out, using the earlier system together with the meanings that have been newly assigned.
3. Finally the new system becomes autonomous. It can be detached, in a sense, from the template that helped to produce it and can acquire meanings and interpretations different from, or more general than, those that were first assigned.

The first of these stages is crucial psychologically because when the child (or adult) first assigns meaning to a new representation, this meaning quickly becomes the "real" meaning for the person. Here we see the power of *initial imagery.*

For example, early experiences with numbers may involve assigning the new symbol (the number) to the result obtained from counting the objects in a concrete set (i.e., reciting the natural numbers, as elements of the set are touched or moved, until the last one is reached). The prior system is the child's internal visual and kinesthetic imagery for representing objects; the new system is the system of numeration. The result of counting becomes what a number "really is" to the child.

One can now work out a lot of arithmetic, using the object-system above as a template: base-ten numeration, arithmetic operations for whole numbers, and so on. This development corresponds to the second stage above. The longer it goes on without alternative interpretations of "number" being considered, the more counterintuitive other interpretations may later seem.

But there are important aspects of number that the object-system does not support. Zero cannot be achieved by counting the elements of an empty set. Negative numbers cannot be understood this way, nor do fractions qualify as numbers. That is why the extension of the meaning of "number" to these entities has a metaphorical aspect. This can be a serious obstacle to later mathematical development. The system of numeration becomes autonomous only to the extent that new meanings and interpretations are offered *beyond* the original ones on which the system has been built.

EXAMPLES: CHILDREN'S REPRESENTATIONS OF NEGATIVE NUMBERS

We shall illustrate some aspects of the foregoing discussion with examples of different representations of negative numbers.

Typically, signed numbers and their manipulation are seen as a topic appropriate for the middle school grades, and in some curricula they are not introduced until the level of prealgebra. This allows plenty of time for cognitive obstacles not only to develop but to become well secured in the representational structures of the children. Students learn early in school that "you can't take away a larger number from a smaller number," and later they must tear down and reconstruct their cognitions.

Davidson (1987), among others, criticized the usual sequence in which children are introduced first to counting numbers, addition, and subtraction with manipulatives, then to computation with written numerals, and finally to negative integers. He suggests this leads to misconceptions at each stage and argues that much earlier teaching of negative numbers with manipulatives is possible. Carraher (1990) conducted clinical interviews with adults and with children in grades 4, 5, and 6, some of whom were receiving instruction about signed numbers in school. The study demonstrated both children's and adults' ability to add directed numbers without previous instruction, with no significant difference related to school instruction. But subjects experienced difficulties solving problems on paper. Carraher suggests "the need to distinguish theoretically between the availability of a semantic representation for negative numbers and the use of the mathematical representation through signs" (p. 229). Peled, Mukhopadhyay, and Resnick (1989) have provided additional evidence that children construct internal representations of negative numbers prior to formal school instruction. For additional discussion, empirical work on negative numbers, and recommendations concerning their teaching, see Aze (1989), Hativa and Cohen (1995), Hefendehl-Hebeker (1991), Hitchcock (1997), and Streefland (1996).

How, then, might a young child of age seven assign initial meaning to the notion of a number "less than zero"? To understand the structure of a developing internal representational system, it is useful to try to observe behaviors attendant on its construction, allowing us to infer that some parts of the system are already being developed and some are not. Two of the possible kinds of meaning assigned for negative numbers are related to the two types of external representations mentioned above: number line representation and signed cardinality representation. Thus we set out to interview children in two different, structured external representational contexts. The interviews took place during the initial phase of an exploratory study in progress that included subsequent group activities with signed numbers.

It is worth taking a moment to comment on the complexity of external representations for signed numbers. We were focusing on the possible meaning of –1 as a *position location* one unit to the left of an arbitrary zero-point or origin of the number line. (Of course, "to the left'" is simply the common convention.) But this assignment of meaning is different from another possible meaning of –1, which is the *operation* of traveling, or the command to travel, one unit to the left on the number line from any starting point. Usually the latter meaning is associated with "subtracting 1" or "adding –1," but note that the 1 that is subtracted or the –1 that is added does not have any immediate connection to the meaning of –1 as a *location.* The different meanings of –1 that are possible here contribute to the skill that is needed in using the number line effectively as a classroom teaching device.

In order to study students' internal representations of signed numbers, we interviewed twelve students from a mixed first- and second-grade classroom in a small suburban private school. The students were of mixed abilities, and none of them had had previous schooling related to negative numbers.

One of the presented representational contexts was related most closely to the number line, embodying *ordinal* relationships. We started with a paper strip consisting of blank circular spaces that we called a "path." Additional spaces could be attached to either end of the path. The strip was placed on the table so that the left end was in front of the student, and the right end (from the student's perspective) hung over the edge of the table (see fig. 1.1).

The student was given a set of cards marked with the numerals from 0 to 7 and additional blank cards. The student was first asked to use the cards to name the spaces on the path. We expected most of the children to come up with the sequence 1, 2, 3, ..., starting from the leftmost space. If another labeling was offered, as often happened, we accepted the answer but asked about other possibilities until we had evoked a solution like this one. This established a convenient representational basis for further questions.

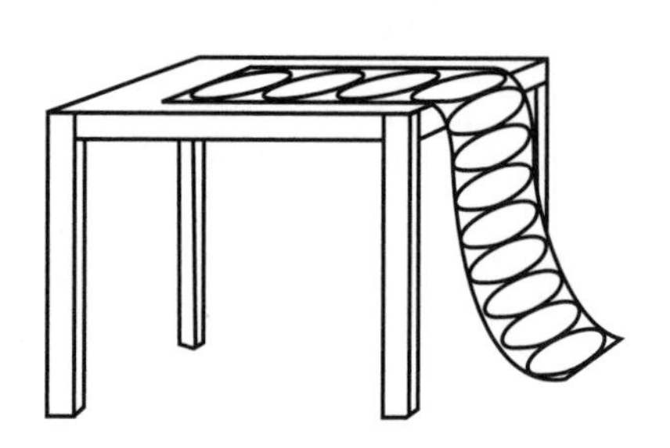

Fig. 1.1. An ordinal representational context for the task-based interviews

At that point the (external) representation (labeled by the counting numbers) was expanded, in an attempt to include zero and negative numbers. A blank space was attached at the end of the path to the left of the space the child had labeled 1. The child was asked to name the new space without changing the way the other spaces were named. If the child named this space with the 0, he or she was asked then to name another new space, which was attached to the left of the one named 0. At each step, the child was asked to explain why.

The second external representational context was related to signed *cardinality*. We introduced a game with a spinner that could point either to a sad (frowning) face or to a smiling face (see fig. 1.2). Someone who spun the arrow so that it ended pointing at the smiling face would gain a point. Otherwise, the player would lose a point. The player started with 2 points at the beginning of the game (to avoid a negative score too early in the interview).

The student was asked to spin the arrow and then tell how the score changed as a result. The part of the interview that followed allowed us to infer whether the student's internal representation could include (or extend to include) zero and negative numbers in this context. The situation where the student started to lose points either happened by itself or was created by the interviewer so that the child faced the situation in which the score was 0 points and the arrow pointed to the sad face.

Fig. 1.2. A signed cardinality representational context for the task-based interviews

Next, the student was asked to create a method of keeping score. For this purpose the use of chips (black and red) was suggested. Black and red chips were available at the table during the "cardinal" portion of the interview.

Each child experienced the two representational contexts, six in the order "ordinal then cardinal" and six in the opposite order, "cardinal then ordinal." There were five children (among the twelve interviewed in this group) whose cognitive obstacles to "numbers less than zero" appeared firm in both representational contexts. The other seven children evidenced some ability to use the idea of negative numbers in at least one of the representational contexts. We shall shortly look at examples drawn from three of these children. Among other sets of interviewees, nineteen of thirty-four children drawn from two second-grade classrooms in a public suburban school evidenced at least partial familiarity with negative numbers, as did all ten students interviewed from a multiage third- and fourth-grade private school classroom. These qualitative observations suggest that many contemporary students begin to develop informal internal representations of negative numbers rather earlier than is widely believed. This allows opportunities for various misconceptions to form.

We observed a variety of behaviors, from which we have concluded that development in either of the two representational contexts can occur independently of the other. Our results for negative numbers are similar to those of Piaget (1952) for the learning of natural numbers, who concluded that cardinal and ordinal concepts of number develop separately and are later integrated. But other researchers have reached different conclusions for natural numbers. Brainerd (1976) reports that ordinal number concepts are mastered earlier than cardinal concepts, with training for ordinal number concepts significantly more successful.

Let us consider some of the interviews (the names of all the children are changed).

Interview 1: Jeremy

Jeremy is a boy in the second grade, seven years and three months old at the time of the interview. He is considered by his teachers to be an average student, and he has not seemed to show particular interest in mathematics.

Jeremy rapidly named the spaces of the path 1, 2, 3, 4, … He volunteered that there was a way to name more spaces, if they were attached to the right end of the path. When asked to name an additional space attached to the left end of the path, without renaming all the rest, he named it 0.

Interviewer: Suppose I put another space here, but I don't want you if possible to move all the others. Could you name this space somehow?

Jeremy: (*without a pause*) Negative one.

Interviewer: Negative one. Can you write this for me?

Jeremy: (*Writes* "–1")

Interviewer: And why did you do it this way?

Jeremy: 'Cause it's the way I learned to do it.

Interviewer: Because it's the way you learned it?

Jeremy: Yeah.

Interviewer: Who taught you?

Jeremy: Ah … No one in particular.

Later, Jeremy said that "negative two goes before negative one." Thus he seemed to demonstrate a stable, well-established representational assignment of meaning to negative numbers in this (ordinal) context, using the words himself spontaneously and writing the conventional symbols. In his affect, he appeared to be confident and secure.

In the second part of the interview, Jeremy wanted to spin the arrow not for himself, but for "Mister X." After Mister X had 1 point, Jeremy spun the arrow for Mister X, and the arrow pointed to the sad face.

Jeremy: Now … zero points.

Interviewer: Because he lost a point, OK?

Jeremy: OK. (*Spins the arrow again, and it points to the sad face again.*)

Still zero points. (*Slowly, in a low voice.*) Or negative one point.

Interviewer: Zero or negative one?

Jeremy: Still zero.

Interviewer: Zero. Why zero?

Jeremy: Because, because it hits right, right here (*points at the sad face*).

Interviewer: So he had zero points, then he hit the frowny face. How many points does he have now?

Jeremy: Zero.

Interviewer: Still zero?

Jeremy: Still zero.

Interviewer: OK.

Jeremy: (*Spins the arrow; the arrow points at the sad face again; Jeremy takes a deep breath.*) Still zero.

Interviewer: Why is it still zero?

Jeremy: Or negative one.

Interviewer: Which one is …

Jeremy: Probably negative one … or zero. You could see it either way. I would say zero.

Interviewer: OK. OK. Why do you prefer zero?

Jeremy: Because most people say it like that.

We see an interesting contrast with the earlier, ordinal context, where Jeremy was confident and seemed to have firmly established a representational significance for negative numbers. Here he hesitated, considered the possibility, but did not yet accept the idea (though it was his own idea) that a score (a cardinal quantity) can be negative. He was not certain, but he preferred zero and thought that "most people say it like that." His affect was serious and no longer confident, as evidenced by his hesitance and his deep breath. Jeremy appears to have completed the first "assignment of meaning" stage with respect to ordinal representation of negative numbers but to have unstable assignments of meaning when the representation is not ordinal. During the task he appeared to be partially building an internal, cardinal representation for negative numbers.

Next, we consider an interview that is almost a mirror image of the one with Jeremy.

Interview 2: James

James is a boy in the second grade, aged seven years, nine months. He is considered by his teachers to be an average student. He likes mathematics.

When asked to name the spaces, James named them in order from left to right, starting with 1. After an additional space was attached at the left end of the path (to the left of 1), James suggested naming it with 0, because "zero goes before one." Then the following happened:

Interviewer: Suppose that I put another circle just in front of this one (*attaches a blank space in front of the one named with 0*).

James: (*Starts sliding the cards with the numbers one space to the left*)

Interviewer: Oh, no, no, wait, wait a little, James, you are too fast. I like this way of naming them. And I would not like to rename all the circles. Could we still have all these names here but name this circle somehow?

James: Um, no. If we move them, we can put the 7 there (*pointing to the last visible space at the right, which presently has 6 in it*).

Interviewer: Yes, I understand. But if we are not going to move them, and 7 will go here (*places 7 to the empty space to the right of the space labeled 6*) and 8 will go here (*points to the next empty space on the child's right*), is there any way to name this shape without renaming anything else?

James: No. If you can't move them.

Interviewer: If I can't move, there is no way?

James: Yes.

Thus James showed no internal representation of negative numbers in the ordinal context. But later something else occurred. (Recall that black and red chips were available on the table during the entire cardinal part of the interview.) Mister X, for whom James was spinning the arrow, had 0 points:

Interviewer: Let's spin again.

James: (*Spins the arrow; the arrow points at the smiling face.*) Now ... one. (*James takes one red chip from the pile of red chips.*) He has one.

Interviewer: Oh, he is keeping track using chips?

James: Yes.

Interviewer: That makes sense. So if he gets one point, he takes a chip.

James: (*spins the arrow; it points at the smiling face again. James takes another red chip.*) He gets another.

Interviewer: Now he has two chips.

James: Um-hum (*spins the arrow, moves back one of his chips*).

Interviewer: Losing chip, losing a point.

James: Um-hum (*spins the arrow, moves away another chip*).

Interviewer: Losing another point.

James: (*Spins again.*) How can I lose a point if I don't have any?

Interviewer: How can you lose a point if you don't have any? So what's …

James: Negative one?

Interviewer: Negative one?

James: Here's negative one. (*James takes a black chip. He smiles.*)

Interviewer: Oh, why did you take this?

James: Because it is not red. So it's negative one.

Interviewer: A-ha, so this is for normal points (*points at red chip*); this is for … (*points at black chip*).

James: Negative points.

Interviewer: OK.

James: (*Spins.*) Now I get … zero. (*First James trades the black chip for the red one, but then he moves away the red one also.*)

In James's interview, it seemed at first that he would show a cognitive obstacle to negative numbers in interacting with the cardinal external representation, as he had in interacting with the ordinal. "How can I lose a point if I don't have any?" he asked. But then he spontaneously proposed using negative numbers to mark a score less than zero. He also created a reasonable way to keep score using chips of two different colors. The way James behaved when he suggested using negative numbers—he looked amused, and acted as if what he was doing was a joke—is indicative of rich affect and suggests that this might even have been the first time James had used negative numbers in such a situation.

Interview 3: Alice

It is interesting to compare these responses with those of Alice, a girl in the second grade, aged seven years and six months. Like James, Alice showed no prior familiarity with negative numbers in the ordinal representation. She appeared to alternate between being willing to use 0 as a label for more than one space and asserting that "there is nothing there" (without a label). In the cardinal representation, Alice also showed no familiarity with negative numbers, but she came up with an interesting suggestion after the game had been played for a while.

Interviewer: Suppose I had zero points, and I spin and get a sad face.

Alice: We have nothing. And then, the next point we get, we would not get it.

Interviewer: Could you please repeat it?

Alice: And since we have one point to take away, the next point we have, we don't get it.

Alice did not assign a label like –1 to the situation. Nevertheless, she invented a rule that incorporates one important structural feature of signed cardinality—the idea of offsetting the lost point with a subsequent gained point. This indicates that even in the absence of any clear notational assignment of meaning, some important conceptual development was taking place.

Comparison

Of course we cannot validly base an elaborate picture of individual students' internal representations on short interview segments. Nevertheless, some possible inferences can be offered for explicit discussion. Let us note briefly the kinds of internal representations the children in these interviews seem to use in interaction with the externally presented tasks as well as the structures of these representations.

In interacting fluently with the ordinal task, Jeremy evidences an internal *spatial* representation of signed numbers extending in both directions in an ordered line, the numbers growing from left to right. His *verbal* descriptions fit such a structure. He also has some *formal notational* representation of negatives, writing "–1" to represent the space to the left of zero that he called "negative one." In contrast, James's internal spatial representation seems to treat the counting numbers as arranged in a rigid order, with zero as the first one. The structure permits these numbers to be moved as a whole group, but each number has a fixed, unchangeable place. Negative numbers are not part of this representation at all (a cognitive obstacle).

However, James spontaneously assigns symbolic meaning to the black chips, as representing scores below zero, and uses this. He thus finds a *strategic* solution to a problem that he represents meaningfully, creating (at least tacitly) a structure that extends to negatives in the cardinal context. His smile is evidence for his affect of satisfaction with this. Alice's internal representation here is especially interesting in comparison with that of James. She also spontaneously develops a partial representation for *operations* with numbers that are effectively less than zero but *without* the act of symbolizing such numbers with notations or chips. Rather, she describes *verbally* a rule she has created for making future score adjustments. And the apparent change in Jeremy's emotional state from confident to unsure suggests that cardinal representation has a more problematic status for him. There is evidence that his operational understanding of (positive) whole numbers is represented verbally, formally, visually, and kinesthetically during the spinner game, but these representations do not extend structurally to negatives. Jeremy's first steps in assigning meaning here are just being taken.

CONCLUSION

We teach mathematics most effectively when we understand the effects on students' learning of external representations and structured mathematical activities. To do this, we need to be able to discuss how students are representing concepts internally—their assignments of meaning, the structural relationships that they develop, and how they connect different representations with one another. This article has surveyed and discussed a number of these ideas. Further examples and discussions of the role of representation in school mathematics will be found throughout this volume.

The research framework presented has allowed us a more detailed look at the meanings that children in the early elementary grades can give to "numbers below zero" before they encounter them in school. There is evidence that structurally different internal representations associated with negative numbers can develop independently of one another, at least in the informal, early stage of meaning. The ability to assign meaning and to use negative numbers in an ordinal external representational context of activity neither implies nor is implied by the presence of these abilities in a cardinal external representational context.

More generally, we have presented evidence of young children's strong representational capabilities. The negative number concept to which we have devoted attention is just one example. Here the age at which effective internal representation becomes possible seems earlier than commonly assumed. This suggests that a much earlier foundation for children's representational ability in this regard could be considered in the school curriculum. We have also seen the early—and perhaps unnecessary—development of cognitive obstacles, which later will need to be overcome. Early attention to children's representations of number, both positive and negative, making use of both ordinal and cardinal external representational contexts, could remove these obstacles before they become long-term impasses, generating strong and positive affect as well.

In teaching every mathematical topic, we should see the development of strong, flexible internal systems of representation in each student as the essential goal. New technology is providing new external representational systems that can help reach this goal. Ultimately, this awareness of representational goals is necessary to achieve the desired universality of access to a high level of achievement in schools.

REFERENCES

Aze, Ian. "Negatives for Little Ones?" *Mathematics in School* 18 (1989): 16–17.

Brainerd, Charles J. "Analysis and Synthesis of Research on Children's Ordinal and Cardinal Number Concepts." In *Number and Measurement,* edited by Richard A.

Lesh and David A. Bradbard, pp. 189–231. Columbus, Ohio: ERIC, 1976. (ERIC Document Reproduction no. ED120 027)

Bruner, Jerome S. *The Process of Education.* Cambridge, Mass.: Harvard University Press, 1960.

———. "The Course of Cognitive Growth." *American Psychologist* 19 (1964): 1–15.

Carraher, Terezinha Nunes. "Negative Numbers without the Minus Sign." In *Proceedings of the 14th International Conference for the Psychology of Mathematics Education,* edited by George Booker, Paul Cobb, and Teresa N. de Mendicuti, Vol. 3, pp. 223–30. Oaxtepex, Mexico: Program Committee of the 14th PME Conference, 1990.

Davidson, Philip M. "How Should Non-Positive Integers Be Introduced in Elementary Mathematics?" In *Proceedings of the 11th International Conference for the Psychology of Mathematics Education,* edited by Jacques C. Bergeron, Nicolas Herscovics, and Carolyn Kieran, Vol. 2, pp. 430–36. Montreal, Que.: Program Committee of the 11th PME Conference, 1987.

Davis, Robert B. "Discovery in the Teaching of Mathematics." In *Learning by Discovery: A Critical Appraisal,* edited by Lee S. Shulman and Evan R. Keislar, pp. 114–28. Chicago: Rand McNally, 1966.

———. *Learning Mathematics: The Cognitive Science Approach to Mathematics Education.* Norwood, N.J.: Ablex, 1984.

Davis, Robert B., and Carolyn A. Maher. "How Students Think: TheRole of Representations." In *Mathematical Reasoning: Analogies, Metaphors, and Images,* edited by Lyn English, pp. 93–115. Mahwah, N.J.: Lawrence Erlbaum Associates, 1997.

DeBellis, Valerie M. *Interactions between Affect and Cognition during Mathematical Problem Solving.* Doctoral diss., Rutgers University. Ann Arbor, Mich.: University Microfilms # 96-30716, 1996.

Dienes, Zoltan P. *Mathematics in the Primary School.* London: Macmillan, 1964.

———. "Some Reflections on Learning Mathematics." In *Learning and the Nature of Mathematics,* edited by William E. Lamon, pp. 49–67. Chicago: Science Research Associates, 1972.

Edwards, Laurie. "Embodying Mathematics and Science: Microworlds as Representations." *Journal of Mathematical Behavior* 17, no. 1 (1998): 53–78.

English, Lyn D. *Mathematical Reasoning: Analogies, Metaphors, and Images.* Mahwah, N.J.: Lawrence Erlbaum Associates, 1997.

Ernest, Paul. *The Philosophy of Mathematics Education.* Basingstoke, Hampshire, U.K.: Falmer Press, 1991.

Even, Ruhama. "Factors Involved in Linking Representations of Functions." *Journal of Mathematical Behavior* 17, no. 1 (1998): 105–21.

Goldin, Gerald A. "Cognitive Representational Systems for Mathematical Problem Solving." In *Problems of Representation in the Teaching and Learning of Mathe-*

matics, edited by Claude Janvier, pp. 125–245. Hillsdale, N.J.: Lawrence Erlbaum Associates, 1987.

———. "Representational Systems, Learning, and Problem Solving in Mathematics." *Journal of Mathematical Behavior* 17, no. 2 (1998a): 137–65.

———. "Retrospective: The PME Working Group on Representations." *Journal of Mathematical Behavior* 17, no. 2 (1998b): 283–301.

———. "Affective Pathways and Representation in Mathematical Problem Solving." *Mathematical Thinking and Learning* 2, no. 3 (2000): 209–19.

Goldin, Gerald A., and Claude Janvier, eds. "Representations and the Psychology of Mathematics Education, Part I." (Special Issue) *Journal of Mathematical Behavior* 17, no. 1 (1998a): 1–134.

———. "Representations and the Psychology of Mathematics Education, Part II." (Special Issue) *Journal of Mathematical Behavior* 17, no. 2 (1998b): 135–301.

Goldin, Gerald A., and James J. Kaput. "A Joint Perspective on the Idea of Representation in Learning and Doing Mathematics." In *Theories of Mathematical Learning*, edited by Leslie P. Steffe, Pearla Nesher, Paul Cobb, Gerald A. Goldin, and Brian Greer, pp. 397–430. Hillsdale, N.J.: Lawrence Erlbaum Associates, 1996.

Hativa, Nira, and Dorit Cohen. "Self Learning of Negative Number Concepts by Lower Division Elementary Students through Solving Computer-Provided Numerical Problems." *Educational Studies in Mathematics* 28 (1995): 401–31.

Hefendehl-Hebeker, Lisa. "Negative Numbers: Obstacles in Their Evolution from Intuitive to Intellectual Constructs." *For the Learning of Mathematics* 11 (1991): 26–32.

Hitchcock, Gavin. "Teaching the Negatives, 1870–1970: A Medley of Models." *For the Learning of Mathematics* 17 (1997): 17–25, 42.

Janvier, Claude, ed. *Problems of Representation in the Teaching and Learning of Mathematics.* Hillsdale, N.J.: Lawrence Erlbaum Associates, 1987.

Kaput, James J. "Linking Representations in the Symbol System of Algebra." In *A Research Agenda for the Teaching and Learning of Algebra*, edited by Carolyn Kieran and Sigrid Wagner, pp. 167–94. Reston, Va.: National Council of Teachers of Mathematics; Hillsdale, N.J.: Lawrence Erlbaum Associates, 1989.

———. "Technology and Mathematics Education." In *Handbook on Research in Mathematics Teaching and Learning*, edited by Douglas Grouws, pp. 515–56. New York: Macmillan, 1992.

———. "The Representational Roles of Technology in Connecting Mathematics with Authentic Experience." In *Didactics of Mathematics as a Scientific Discipline*, edited by Rolf Biehler, Roland W. Scholz, Rudolf Straesser, and Bernard Winkelmann. Dordrecht, Netherlands: Kluwer, 1993.

Kosslyn, Stephen M. *Image and Mind.* Cambridge, Mass.: Harvard University Press, 1980.

Lesh, Richard, Marsha Landau, and Eric Hamilton. "Conceptual Models in Applied Mathematical Problem Solving Research." In *Acquisition of Mathematics Concepts and Processes,* edited by Richard Lesh and Marsha Landau, pp. 263–343. New York: Academic Press, 1983.

McLeod, Douglas B., and Verna M. Adams, eds. *Affect and Mathematical Problem Solving: A NewPerspective,* New York: Springer-Verlag, 1989.

Mesquita, Ana Lobo. "On Conceptual Obstacles Linked with External Representation in Geometry." *Journal of Mathematical Behavior* 17 (1998): 183–95.

Montessori, Maria. *La scoperto del bambino.* 6th ed. Milan, Italy: Garzanti, 1962. (English translation: *The Discovery of the Child.* Translated by M. Joseph Costelloe. New York: Ballantine Books, 1972.)

———. *Basic Ideas of Montessori's Educational Theory: Extracts from Maria Montessori's Writing and Teaching.* Oxford, England: Clio Press, 1997.

National Council of Teachers of Mathematics (NCTM). *Principles and Standards for School Mathematics.* Reston, Va.: NCTM, 2000.

Palmer, S. E. "Fundamental Aspects of Cognitive Representation." In *Cognition and Categorization,* edited by Eleanor Rosch and Barbara B. Lloyd. Hillsdale, N.J.: Lawrence Erlbaum Associates, 1977.

Peled, Irit, Swapna Mukhopadhyay, and Lauren B. Resnick. "Formal and Informal Sources of Mental Models for Negative Numbers." In *Proceedings of the 13th International Conference for the Psychology of Mathematics Education,* edited by Gerard Vergnaud, Janine Rogalski, and Michele Artigue, Vol. 3, pp. 106–10. Paris: G. R. Didactique, Laboratoire de Psychologie du Développement et de l'Education de l'Enfant, 1989.

Piaget, Jean. *The Child's Conception of Number.* New York: Humanities Press, 1952.

———. *Science of Education and the Psychology of the Child.* New York: Viking Press, 1969.

Polya, George. *Induction and Analogy in Mathematics.* Princeton, N.J.: Princeton University Press, 1954a.

———. *Patterns of Plausible Inference.* Princeton, N.J.: Princeton University Press, 1954b.

Presmeg, Norma. "Prototypes, Metaphors, Metonymies and Imaginative Rationality in High School Mathematics." In *Educational Studies in Mathematics* 23 (1992): 595–610.

———. "Metaphoric and Metonymic Signification in Mathematics." In *Journal of Mathematical Behavior* 17 (1998): 25–32.

Skinner, B. F. *Science and Human Behavior.* New York: Free Press, 1953.

———. *About Behaviorism.* New York: Alfred A. Knopf, 1974.

Streefland, Leen. "Negative Numbers: Reflections of a Learning Researcher." *Journal of Mathematical Behavior* 15 (1996): 55–77.

Sund, Robert B., and Anthony J. Picard. *Behavioral Objectives and Evaluational Measures: Science and Mathematics.* Columbus, Ohio: Charles E. Merrill Publishing Co., 1972.

von Glasersfeld, Ernst, ed. *Radical Constructivism in Mathematics Education.* Dordrecht, Netherlands: Kluwer, 1991.

2

Representation and Abstraction in Young Children's Numerical Reasoning

Constance Kamii

Lynn Kirkland

Barbara A. Lewis

ARITHMETIC instruction for young children usually begins with concrete objects and progresses to the use of "semiconcrete" aids, such as pictures, and finally to "abstract" symbols. The assumption is that these representations, when properly introduced, convey the intended mathematical meanings to children. Piaget's research shows, however, that children's understanding does not depend on representations but on children's level of abstraction. To explain why this is so, we begin by describing the distinction Piaget made between *representation* and *abstraction.*

THE RELATIONSHIP BETWEEN REPRESENTATION AND ABSTRACTION

Piaget (1951) distinguished between two kinds of tools we use in representation—*symbols,* such as pictures and tally marks, and *signs,* such as spoken words and written numerals. He used the term *symbol* differently from common parlance.

In Piaget's theory, "symbols" bear a resemblance to the objects represented and can be invented by each child. For example, children can draw eight apples without any instruction. They can likewise use eight fingers, eight counters, or eight tally marks as symbols without being shown how to do this. (Eight tally marks resemble eight apples, but the numeral "8" does not.)

Examples of "signs" are the spoken words *apple* and *eight* and the written numeral "8." Signs do not resemble the objects represented, and their source is conventions, which are made by people. In other words, unlike pictures and tally marks, signs cannot be invented by children. Signs are integral parts of systems and require social transmission. Other examples of conventional systems are mathematical signs (such as "+"), musical notations, and the Morse code.

To clarify the relationship between representation and abstraction, it is necessary to discuss the three kinds of knowledge Piaget (1971, 1951) distinguished according to their ultimate sources and the two types of abstraction involved in the acquisition of each kind of knowledge.

Three Kinds of Knowledge

The three kinds of knowledge are physical, social (conventional), and logico-mathematical knowledge. *Physical knowledge* is the knowledge of objects in external reality. The color and weight of counters or any other object are examples of physical knowledge. The fact that counters do not roll away like marbles is also an example of physical knowledge. The ultimate source of physical knowledge is thus partly *in* objects, and physical knowledge can be acquired empirically through observation. (Our reason for saying "partly" will be explained shortly.)

Social knowledge is the knowledge of conventions, which were created by people. Examples of social knowledge are languages such as English and Spanish and holidays such as the Fourth of July. The ultimate source of social knowledge is thus partly in conventions. (Our reason for saying "partly" will also be clarified shortly.)

Logico-mathematical knowledge consists of mental relationships, and the ultimate source of these relationships is the human mind. For instance, when our ancestors (as well as each one of us) saw two pebbles on the ground, they could think about them as being *different* or *similar*. It is just as true to say that the pebbles are different (because pebbles on the ground are seldom identical) as it is to say that they are similar (because they share common characteristics). The similarity and difference exist neither *in* one pebble nor *in* the other.

Another relationship an individual can create between the pebbles is *two*. The pebbles can be observed empirically, but the number "two" cannot. The ultimate source of logico-mathematical knowledge is each person's mind.

Piaget conceptualized two kinds of abstraction—empirical abstraction and constructive abstraction. ("Constructive abstraction" is also known as "reflective" or "reflecting" abstraction. The French term Piaget usually used was *abstraction réfléchissante*, which has been translated to "reflective" or "reflecting" abstraction. He also occasionally used the term *constructive abstraction*, which seems easier to understand.) This distinction helps to differentiate external and internal sources of knowledge.

Empirical and Constructive Abstraction

In *empirical abstraction,* we focus on certain properties of an object and ignore the others. For example, when we abstract the color of an object (physical knowledge), we simply ignore the other properties such as weight and the material with which the object is made (plastic, for instance). These properties are knowable empirically through our senses, and we choose only the one(s) we want to abstract.

In contrast, *constructive abstraction* involves the *making* of *mental* relationships between and among objects. Relationships such as "similar," "different," and "two" (logico-mathematical knowledge) are made by constructive abstraction. Whereas properties of objects are abstracted *from objects,* relationships are abstracted *from our mental actions* (thinking) on objects.

Having made the theoretical distinction between empirical and constructive abstraction, Piaget went on to say that in the psychological reality of the child, one cannot take place without the other. For example, we could not construct the relationship "different" if all the objects in the world were identical. Similarly, the relationship "two" would be impossible to create if children thought that objects behave like drops of water, which can combine to become one drop.

Conversely, we could not construct physical knowledge, such as the knowledge of "red," if we did not have the category of "color" (as opposed to every other property) and the category of "red" (as opposed to every other color). A logico-mathematical framework (built by constructive abstraction) is necessary for empirical abstraction because children could not "read" facts from external reality if each fact were an isolated bit of knowledge unrelated to the knowledge already built and organized. This is why we said earlier that the source of physical knowledge is only *partly* in objects and that the source of social knowledge is only *partly* in conventions.

Although constructive abstraction cannot take place independently of empirical abstraction up to about age six, it later becomes possible for constructive abstraction to take place independently. For example, once the child has constructed number, he or she becomes able to put these relationships into relationships without empirical abstraction. By putting four "twos" into relationships, for instance, children become able to deduce that $2 + 2 + 2 + 2 = (2 + 2) + (2 + 2)$, that $4 \times 2 = 8$, and that if $4x = 8$, x must be 2.

The Conservation-of-Number Task

The conservation-of-number task clarifies constructive abstraction. This task is sketched briefly, and the reader can find further details in Kamii (1982, 1985, 2000). Most four- and five-year-olds make a one-to-one correspondence when asked to put out the same number of counters as the eight that the interviewer has aligned. After two rows of eight counters have been made, the interviewer tells the child to "watch carefully what I'm going to do." The interviewer then spreads out the counters in one row and pushes together those in the other row.

The child is now asked the conservation questions: "Are there just as many here (row A) as here (row B), or are there more here (row A) or more here (row B)?" and "How do you know?" At age four and slightly beyond, most children say that one row has more because it is longer.

Conservation is attained between five and six years of age by most middle-class children, who say that there are just as many in row A as in row B. When asked, "How do you know?" conservers give one of the following three logical arguments:

- "You didn't add or take anything away" (the *identity* argument).
- "If we put the chips back to the way they were before, you'll see that the number is the same" (the *reversibility* argument).
- "One row is longer, but there's more space between the chips" (the *compensation* argument).

The conservation task is a test of children's logic. Counters are objects that can be observed empirically (physical knowledge), but observation is not enough for children to deduce *logically* that the quantity in the two rows remains the same. Only when children can put the objects into numerical relationships (by constructive abstraction) can they deduce *with the force of logical necessity* that the two rows *must* have the same number.

The reader may be wondering why nonconservers do not try counting. When children begin to count, we usually ask them if they can answer the question without counting. When children's logic is well developed, conservation becomes obvious, without any need to resort to an empirical procedure like counting.

Note that the conservation task is given with concrete objects. The counters are empirically observable (physical knowledge), but *number* (logico-mathematical knowledge) is not. In other words, number is constructed by each child, through constructive abstraction, and is *always* abstract.

Young Children's Graphic Representation of Groups of Objects

Sinclair, Siegrist, and Sinclair (1983) individually interviewed four-, five-, and six-year-olds in a kindergarten and day-care center in Geneva, Switzerland, using up to eight identical objects, such as pencils and small rubber balls. No formal, academic instruction was given at that time in public kindergartens in Geneva.

Presenting the child with three small rubber balls, for example, as well as a pencil and paper, the interviewer asked, "Could you put down what is on the table?" This request was carefully worded to avoid using terms such as "how many" and "number" that would have suggested quantification.

After making several similar requests (with two balls, then five houses, for example), the researchers asked the child, "Could you write 'three' (then 'four,' and so on)?" These requests were made to find out if the child could write numerals when explicitly asked to do so in the absence of objects.

The following six types of notations were found (see fig. 2.1):

1. *Global representation of quantity.* Examples of this type are "////" for three balls and "/////" for two balls. These children can be said to be representing the vague quantitative idea of "a bunch" or "more than one."

Types	Three balls	Two balls	Five houses
1.	/\\\	/|\\\\	/|\V/
2.	8	B	5
3.	PPP	PP	77777
3.	TIL	JT	IJTTI
3.	AEI	OI	9AƎOI
4.	12Ɛ	12	12Ɛ42
4.	333	22	55555
5.	3 TRO	2 DƎ	5 siñ
6.	4 crèion	deu bal	Ƨ mèzone

Fig. 2.1

2. *Representation of the object-kind.* Type 2 notations show a focus on the qualitative rather than the quantitative aspect of each set. The examples in figure 2.1 are a "B" for three balls and two balls and the drawing of a house for five houses.

3. *One-to-one correspondence with symbols* ("symbols" in the Piagetian sense). Some children invented symbols to represent the correct number, and others used conventional letters such as "TIL" for three balls. This is the first type in which precise numerical ideas made an appearance.

4. *One-to-one correspondence with numerals.* One of the examples for three balls is "123," and another example is "333." It can be said that the children who wrote these numerals felt the need to represent each object or their action of counting.

5. *Cardinal value alone.* We finally see "3" for three balls and "5" for five houses (along with "invented" spelling in French for the spoken numerals *trois, deux,* and *cinq*).

6. *Cardinal value and object-kind.* Examples of this type are "4 pencils" and "5 houses." These representations show a simultaneous focus on the quantitative and qualitative aspects of each set.

Type 1 was found mainly among the four-year-olds, and types 5 and 6 were found mostly among those older than five and a half. Types 3 and 4 (one-to-one correspondence) were most frequently found in the middle of the age range, at about the time children become conservers. It must be noted that there are no clear-cut levels in this development, since half of the children used more than one type of representation.

A significant finding is that many children who used only types 1, 2, or 3 were perfectly able to write "3," "4," and so on, when explicitly asked, "Can you write 'three,' (then 'four' and so on)?" The question that arises is "Why did they not write the numerals they knew?"

Our answer to this question is that when children represent reality, they represent *their ideas* about reality and not reality itself (Piaget 1977). When they saw three balls, for example, some children thought about them as "a bunch" and made a type 1 representation. Others thought "balls" and made a type 2 representation. These four-year-olds thought about the objects either from a vaguely quantitative or from a qualitative point of view.

At age five, when children construct number (by constructive abstraction), they tend to make type 3 and type 4 representations. These children think about three objects with numerical precision but still think about each object. Type 4 is especially informative because it shows that even when they use their social knowledge of written numerals, children use them at their respective levels of constructive abstraction. No one teaches children to write "123" or "333" to represent three objects, but types 3 and 4 reveal children's attention to *each object* rather than to the *total quantity.*

Type 5 representation was made mostly by the oldest children, reflecting their thinking about the *total quantity* of objects. At this point, one numeral seems best to them to represent a higher-order unit.

This study illustrates the relationship between abstraction and representation. When children are presented with three balls and their concept of quantity is not yet precise, they represent their vague idea with type 1 symbols. When their idea becomes more precise and numerical (through constructive abstaction), they make types 3–6 representations. Types 3–6 show that children use conventional signs to represent *their concepts of number* at *their level* of constructive abstraction. If they are still thinking about each object in the set, or each counting-word they write "TIL," "123," or "333."

First Graders' Use of Signs to Represent Addition

The discussion so far dealt with children's representation of groups of objects. We now go on to their representation of an operation such as addition. One of us individually interviewed 204 first graders during the second half of the school year in six public schools (two in Japan and four in the

United States—in Chicago and in a small town and a suburb in Alabama). The children had all been completing worksheets for months (since kindergarten in many cases) and could read, write, and define mathematical symbols such as "+" and "=."

The interviewer showed a doll to each child and said, "I'm going to turn him around (turning the doll's back toward the child) so that he won't be able to see what I'm going to do. I'm going to do something here (showing about forty chips and a small, transparent, plastic container), and when I have finished, I want you to write with numbers what I will have done, so that the doll will be able to read your writing and know what happened." The interviewer then dropped three chips simultaneously into the container saying, "First, I am putting three in." She went on to drop two more chips into the container saying, "I am *adding* two." The child was asked to "write with numbers what I just did so the doll will be able to read your writing and know what happened."

Many children wrote "5." When this happened, the interviewer said, "Yes, that's how many there are now. But I first put three in." After emptying the container, the interviewer dropped three chips into it and asked, "Could you first write 'three' on your paper to say that I first put three in?" After the child wrote "3," the interviewer demonstrated with chips again, saying, "I am *adding* two," emphasizing the word *adding.*

The findings are summarized in table 2.1. It can be seen in this table that the children made a wide variety of representations. All the first graders were thoroughly familiar with equations, but only 68 percent, 18 percent, 22 percent, and 62 percent of the four groups, respectively, wrote conventional expressions such as "3 + 2 = 5" and "3 + 2."

Many first graders wrote only two numerals in "3 + 2," "3 + 2 =," "3 2," or "3 5." A 3 and a 2, representing the two original wholes, were written by 18 percent (8 + 2 + 8), 24 percent (0 + 0 + 24), 46 percent (11 + 0 + 35), and 36 percent (14 + 5 + 17), respectively, of the four groups. A 3 and a 5, representing the original whole and the whole at the end, were written by 6 percent, 29 percent, 11 percent, and 9 percent, respectively, of the four groups.

An important point concerns the rarity of the use of the "=" sign in table 2.1. Also, many children wrote "3 2" without the "+" sign. Some children wrote the "+" sign without writing the "=" sign, but no one wrote "=" without writing "+." The "=" sign appears later than the "+" sign because the relationship among 3, 2, and 5 involves a hierarchical relationship, which is very hard for young children to make. When we add two numbers, we combine two wholes (3 and 2) to make a higher-order whole (5), in which the previous wholes become parts. By contrast, the relationship "+" between the two original wholes (3 + 2) does not involve a hierarchical relationship.

TABLE 2.1
*Percents of First Graders in Four Locations Representing the Addition of 3 and 2**

	United States			Japan
	Suburb	Small town	Chicago	
	$n = 62$	$n = 49$	$n = 29$	$n = 64$
3 + 2 = 5	60	18	11	48
3 + 2	8	0	11	14
3 + 2 =	2	0	0	5
32 or $\begin{matrix}3\\2\end{matrix}$	8	24	35	17
35 or $\begin{matrix}3\\5\end{matrix}$	6	29	11	9
Others	18	28	33	6

(The following examples of "Others" are from the United States, since the Japanese sample had only 6% in this category.)

3 + 2 5	+ 3 2 – 5	3 + 5 = 5
3 2 5	$\begin{array}{r}3\\ \hline +2\end{array} = 5$	3 4 5
5 3 2		12345
$\begin{matrix}2\ 3\\5\end{matrix}$	$\begin{array}{c}3\\5\\ \hline\end{array}$	5322

*When a child represented the addition vertically and conventionally, he or she was asked to write "the same thing" horizontally.

Four-year-olds' reaction to the class-inclusion task (Inhelder and Piaget 1964) explains young children's difficulty in making hierarchical relationships. In this task, the child is presented with six miniature dogs and two cats of the same size, for example. He or she is first asked, "What do you see?" so that the interviewer can use words from the child's vocabulary. The child is then asked to show "*all* the animals," "*all* the dogs," and "*all* the cats" with the words from his or her vocabulary (e.g., "doggy"). Only after ascertaining the child's knowledge of these words does the adult ask the following class-inclusion question: "Are there more dogs or more animals?"

Four-year-olds typically answer, "More dogs," whereon the adult asks, "Than what?" The four-year-old's answer is "Than cats." In other words, the question the interviewer asks is "Are there more dogs or more animals?" but what young children *hear* is "Are there more dogs or more cats?" Young children hear a question that is different from the one the adult asks because once they mentally cut the whole (animals) into two parts (dogs and cats), the only thing they can think about is the two parts. For them at that moment, the whole does not exist any more. They can think about the whole, but not when they are thinking about the parts. To compare the whole with a part, the child has to perform two opposite mental actions *at the same time*—cut the whole into two parts and put the parts back together into a whole. This is precisely what four-year-olds cannot do.

The question four-year-olds *hear* is an example of how people understand spoken words. If we can make a part-whole relationship, we understand the term *animals* in the way the interviewer used it. If, however, we can make only a part-part relationship, the only "animals" we can think of besides the dogs are the cats. Young children cannot *see* "the animals" that are in front of their eyes when they cannot *think about* them.

By seven to eight years of age, most children's thought becomes flexible enough to be reversible. *Reversibility* refers to the ability to mentally perform two opposite actions *simultaneously*—in this instance, separating the whole into two parts and reuniting the parts into a whole. In physical, material action, it is impossible to do two opposite things simultaneously. In our minds, however, this is possible when thought (constructive abstraction) has become flexible enough to be reversible. Only when the parts can be reunited in the mind can a child "see" that there are more animals than dogs.

The infrequent use of the "=" sign and of three numerals (3, 2, and 5) reported in table 2.1, as well as in Allardice (1977) and Kamii (1985), can now be explained as a manifestation of young children's difficulty in making hierarchical, part-whole relationships. Children cannot represent (externalize) a part-whole relationship that does not exist in their minds.

EDUCATIONAL IMPLICATIONS

Although we often say that a picture of a dog represents a dog, pictures by themselves do not represent. We also often say that the spoken word *eight* or the written "8" represents a number, but numerals never act by themselves. Representation is what people do. Seven-year-olds at a high level of constructive abstraction represent high-level, logico-mathematical knowledge (a hierarchical relationship) to themselves when they write "3 + 2 = 5," for example. A child at a lower level of abstraction is likely to write "3 2" or "3 5." (As stated earlier, "3 5" is a representation of a low level of abstraction because the child explains that there were three chips [the whole at the beginning] and then five [the whole at the end].)

The educational implication of this theory is that numerical reasoning is fostered when educators focus their efforts on children's constructive abstraction (children's thinking) rather than on representation. If the children's level of abstraction is high, a high level of representation will follow. Textbooks and workbooks overemphasize the writing of mathematical signs (representation at a low level of abstraction) and underemphasize the children's process of thinking.

Georgia DeClark, the first-grade teacher with whom one of us wrote *Young Children Reinvent Arithmetic* (Kamii 1985), played mathematics games with her students every day instead of giving them worksheets. Worried about the effect of not giving worksheets to her students, she gave four of them during the year just to find out what her students could do. She found out each time that her students could quickly write correct answers and that the only children who could not complete the worksheets were those who could not play the games. In other words, children who knew sums (constructive abstraction) could easily write them (representation). Children think harder about numbers (constructive abstraction) while playing games because, unlike worksheets, games are very important *to them.*

Based on the belief that children need these "semiconcrete" materials, workbooks for young children are full of pictures. However, the pictures in workbooks are totally unnecessary, and children who have never used a workbook can draw their own pictures or use their fingers as symbols to solve word problems.

In figure 2.2 are three examples of first graders' responses to the question "How many feet are there in your house?" Figure 2.2a was drawn by a child at a relatively low level of constructive abstraction and has elements of Sinclair and her colleague's types 2 and 3 representations. This child represented not only her knowledge of number but also her physical knowledge of people's arms and heads and her social knowledge of their names. Figure 2.2b still includes much physical knowledge, but only the body parts relevant to the question are represented. Figure 2.2c is a type 3 representation in which the child's logico-mathematical knowledge predominates.

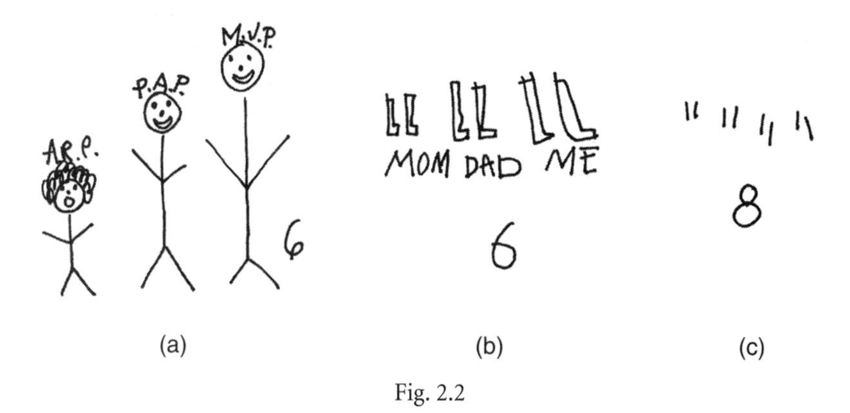

Fig. 2.2

All the drawings in figure 2.2 were made in September, when no one in the class used the numerals they knew as tools for numerical reasoning. By the winter break, many children wrote numerals and drew tally marks as their constructive abstraction became more elaborate. By May, as can be seen in a videotape (Kamii and Clark 2000), those at an advanced level went on to use only numerals to divide 62 by 5 and wrote "5 10 15 20 25 … 60."

Olivier, Murray, and Human (1991) stated that when paper, pencils, and counters are available in classrooms, young children seldom use counters to solve word problems. Rather, they prefer to make drawings like those in figure 2.2. Many teachers we work with in Japan and various parts of the United States have made the same observation and have reported the children's preference for using their fingers and drawings. Counters have properties of their own that interfere with children's representation of *their* ideas. Children prefer to make their own drawings because they can think better with the symbols *they* make by externalizing their own ideas.

REFERENCES

Allardice, Barbara S. *The Development of Representational Skills for Some Mathematical Concepts.* Doctoral diss., Cornell University. Ann Arbor, Mich.: University Microfilms International, 1977.

Inhelder, Bärbel, and Jean Piaget. *The Early Growth of Logic in the Child.* 1959. Reprint, New York: Harper & Row, 1964.

Kamii, Constance. *Number in Preschool and Kindergarten.* Washington, D.C.: National Association for the Education of Young Children, 1982.

———. *Young Children Reinvent Arithmetic.* New York: Teachers College Press, 1985.

———. *Young Children Reinvent Arithmetic.* 2nd ed. New York: Teachers College Press, 2000.

Kamii, Constance, and Faye B. Clark. *First Graders Dividing 62 by 5.* Videotape. New York: Teachers College Press, 2000.

Olivier, Alwyn, Hanlie Murray, and Piet Human. "Children's Solution Strategies for Division Problems." In *Proceedings of the 13th Annual Meeting,* International Group for the Psychology of Mathematics Education, North American Chapter, edited by Robert G. Underhill, Vol. 2, pp. 15–21. Blacksburg, Va.: Virginia Polytechnic Institute, 1991.

Piaget, Jean. *Play, Dreams, and Imitation in Childhood.* 1945. Reprint, New York: W. W. Norton and Co., 1951.

———. *Biology and Knowledge.* 1967. Reprint, Chicago: University of Chicago Press, 1971.

———. *Piaget on Piaget.* Film. New Haven, Conn.: Yale University Media Design Studio, 1977.

Sinclair, Anne, François Siegrist, and Hermina Sinclair. "Young Children's Ideas about the Written Number System." In *The Acquisition of Symbolic Skills,* edited by Don Rogers and John A. Sloboda, pp. 535–41. New York: Plenum Publishing Corp., 1983.

3

Algebra: What Are We Teaching?

Mark Saul

WHAT are we teaching when we think that our students are learning algebra? Algebra is certainly a representational system. But what does it represent? How can we give students access to the concepts represented by algebraic notation?

To explore answers to these questions, we will here borrow a methodology from the medical profession. When a medical researcher wants to know how a certain chemical or physiological process works, he or she often looks at the pathological cases, finding a patient for whom the process works exceptionally well or extremely poorly. In just this way, we can learn much from our least able students.

MEET BARRY

Barry graduated from high school this year. A lanky young man with severe learning disabilities, he spent most of his time in school trying to hide in the back of the class. For the first two weeks of our acquaintance, I didn't know what his voice sounded like. The more quiet he kept, the less he had to deal with the pain of learning.

Most of Barry's four years of high school mathematics was learned in a special class that I taught. By the end of those years, we had become friends. Barry shared with me his enthusiasm for basketball and his quite unexpected sense of humor. He began to take some risks in his personal behavior, although he still never asked questions and would not seek out my attention—even when I was the only other person in the room.

Given these personal and intellectual handicaps, we must consider Barry's learning anything at all about mathematics a success. What has he learned? He can perform the usual arithmetic operations on any two integers. This he finished learning some time toward the end of his sophomore year. Teaching him how to subtract signed numbers was particularly difficult. He had no internal representation for these symbols. We spent a long time creating one, mostly by exploring numerous physical models. As always, Barry had to be pried loose from clinging to a few memorized rules, which he would use in specific situations.

Barry can reduce fractions to lowest terms, but he often doesn't realize that the new fraction and the original represent the same real number. He has trouble comparing two fractions for size unless he uses a calculator to convert both fractions to decimals. By applying algorithms he has memorized, he can perform the usual arithmetic operations on fractions. That is, Barry has some internal representation of signed and rational numbers, but he has not developed a full repertory of such representations, so he sometimes slips into the manipulation of an external representation—a symbol to which he has not attached meaning.

Still, Barry gained enough facility to pass his Regents Competency Examination and earn a high school diploma. Part of Barry's problem is that he was rewarded by passing such an examination, without understanding the mathematics behind it. The emphasis in some of his learning has been on rote procedures, which serve him well in an examination context.

WHAT BARRY DOES NOT KNOW

During his second year of high school, I started working with Barry on algebra, a painful process for both of us. Indeed, neither of us knew what we were talking about. Barry had to learn what algebra did, and I had to learn what algebra meant.

Barry understood the notion of a variable: that it represents an element of a replacement set. He could substitute numbers into a formula and (usually) get the right answer. When he made errors, he understood and corrected them.

Barry solved simple linear equations, with integer solutions, by trial and error. That is, if I gave him the equation $2x + 1 = 7$, he could estimate that the number couldn't be too big, and try $x = 2$, $x = 3$, and so forth, until he got the answer. Indeed, the better he became at this, the better he could thwart my efforts to move him on to a more sophisticated understanding. And I think he knew this.

Barry somehow avoided operating on equations the way most students of algebra do. That is, I couldn"t get him to start with the equation $2x + 1 = 7$, subtract 1 from each side, then divide each side by 2 to get the answer. Why not? This question tormented me almost as much as I tormented Barry.

MEET MERRY

Merry graduated from high school three years ago. She also has significant problems learning but has gotten further than Barry. After two years in a self-contained classroom, she was mainstreamed into a regular mathematics class, two years behind her grade level. This was, for Merry, a major achievement.

Where Barry was taciturn, Merry was voluble. On her good days, she would light up the classroom with her enthusiasm. On her bad days, everyone had to attend to Merry's excited conversation, which became a significant obstacle to learning for others.

Merry initially faced many of Barry's difficulties: a tendency to learn by rote, a shallow grasp of operations on rational numbers, a method of solving equations by trial and error, and a quick reach for the calculator before she had decided what to do with it.

She learned to handle linear equations by "operating on both sides," when they were posed in the form $ax + b = c$. (She had more difficulty if these were in some other form.) She could solve quadratic equations by factoring (if the coefficient of x^2 was 1) or by using the formula (with some difficulty if the radical had to be reduced or the fraction put in lowest terms). She had trouble putting a quadratic equation in canonical form, and she had more trouble adding algebraic fractions unless both denominators were the same simple monomial.

Merry was not fluent in the language of algebra. Although she could factor $3x + 3y$, the factoriziation of $3x + 7x$ gave her pause, and $3x + 7x^2$ was difficult. For Merry, these were three different kinds of problems. Similarly, she could add $1/3 + 4/7$, but not $1/x + 4/y$.

Merry knew how to factor $x^2 - y^2$. She learned to apply this knowledge to factor a number like 2499 as $50^2 - 1^2$. But when asked to factor $2ax^2 - 18ay^2$ or $16x^2y^2 - 64a^2$ or $(x + y)^2 - (x + 3)^2$, she would say that she hadn't learned this yet and would stare at the problem uncomprehendingly.

What did Merry know that Barry did not know? What did Merry herself not know? What did the symbols of algebra represent for each of these students? I needed some time and a certain amount of help to figure this all out.

WHAT ALGEBRA IS NOT

Everyone seems to recognize algebra when they see it: it contains letters, usually drawn from the end of the alphabet. It is obvious to us how this use of the letters differs from their use in language. Letters stand for numbers, they can be added and multiplied, their values can be changed at will, and so on. Before we seek a framework for these intuitive ideas, let us look at what algebra does not consist in. Algebra does not consist merely in the use of variables. For example, a second-grade teacher can ask his students 5 + what? = 7, or 5 + □ = 7. Most of his students will be able to answer. But are they using algebra, in any sense? Not usually. They are more likely to be thinking "5 + 1 = 6, 5 + 2 = 7, so 'box' equals 7."

It is important to note that □ is indeed a variable. The second-grade student knows that □ can stand for any one of a set of numbers. Although she might say that 2 "can be in the □" because it makes the number sentence right, whereas 3 "cannot be in the □" because it makes the sentence wrong, she is still able to conceive of trying to put 3 in the □. She would never put "elephant" in the □: the variable has a definite replacement set in her mind.

Algebra also does not consist in the study of functions. There are many types of functions, and we sometimes use algebra to describe some of them. But we also use geometry (graphs), other diagrams, or natural language. The study of functions is an important application of algebra. The use of algebraic expressions as notation for certain functions can make the algebra come alive, just as the interpretation of "7" as "7 doughnuts" makes that concept more vivid. But algebraic expressions can be used without describing functions, and functions can be expressed without using algebra.

So algebra is not characterized by the use of variables or by the study of functions. This helps explain Barry's difficulties. He had some understanding of the concept of variable, but not of algebra. That is, he knew that $2x + 1 = 7$ is true if $x = 3$ and false if $x = 4$. These are statements about numbers. He could even say that as x increased, so did $2x + 1$. This is a statement about a function. But Barry's understanding of variables and functions did not allow him to solve an equation algebraically. Indeed, he used his understanding of numbers and functions as substitutes for an understanding of algebra.

THREE WAYS TO UNDERSTAND ALGEBRA

We can distinguish three levels of learning about algebra. The first level is algebra as a generalization of arithmetic, the *arithmeticae universalis.* This Latin phrase is that of none other than Sir Isaac Newton, who used it to describe algebra in the elementary textbook that he began on the subject. But perhaps the master should speak for himself (Whiteside 1972, p. 539):

> Common arithmetic and algebra rest on the same computational foundations and are directed to the same end. But whereas arithmetic treats questions in a definite, particular way, algebra does so in an indefinite universal manner, with the result that almost all pronouncements which are made in this style of computation—and its conclusions especially—may be called theorems. However, algebra most excels, in contrast with arithmetic where questions are solved merely by progressing from given quantities to those sought, in that for the most part it regresses from the sought quantities, treated as given, to those given, as though

> they were the ones sought, so as at length and in any manner to attain some conclusion—that is, equation—from which it is permissible to derive the quantity sought. In this fashion the most difficult problems are accomplished, ones whose solution it would be useless to seek of arithmetic alone. Yet arithmetic is so instrumental to algebra in all its operations that they seem jointly to constitute but a unique, complete computing science, and for that reason I shall explain both together.

There are several components to Newton's observations. Perhaps the most important one to the present discussion is Newton's view that algebraic statements are generalized formations of particular arithmetic statements. For example, in Newton's conception, the identity $a^2 - b^2 = (a + b)(a - b)$ has just the meaning that it is true whenever numbers are substituted for a and b.

Algebra considered as a generalization of arithmetic need not contain variables. For example, if we let $b = 1$ in the identity above, we could write the identity as "One less than the square of a number is equal to the product of one more than the number and one less than the number." Indeed, Euclid's algebraic propositions are stated thus—and proved geometrically. Although it is a historical commonplace that Greek algebra did not get nearly as far as Greek geometry, it is interesting to note how far Euclid did get with his words and pictures.

Both Barry and Merry understand this aspect of algebra, although Barry in particular has some trouble using what he knows. This is because he hasn't taken the next step.

We can begin to talk about this next step if we view Newton's remark historically. He and his contemporaries had just emerged from the late Renaissance, and the algebra they knew centered on some remarkable results concerning the solution of polynomial equations. This approach to algebra continued to be fruitful right through the next century or two, but it began to undergo an interesting metamorphosis.

The challenge of finding an algebraic solution to the general quintic equation led to the study of formulas as objects in their own right. Slowly there emerged, from the study of polynomial equations, the study of permutations, then the study of groups of permutations and of fields of numbers. Later years witnessed still greater abstraction of these concepts. The more modern study of algebra has its origins in the study of binary operations on sets.

Today's algebra grew from several roots. One of these was a study of objects other than numbers (symmetries, permutations, and so forth) for which binary operations were defined and for which these operations determined a particular structure. Another was a shift away from the study of particular solutions of equations and toward a study of whole systems of objects (such as fields of numbers) in the context of which these particular solutions could be studied more naturally. This, in turn, led to investigations comparing the properties of these new structures to those of ordinary (real) numbers. Which were prime? Which had inverses? And so on. Today we sometimes look at the "abstract" algebra of groups, rings, and fields as somehow different from the elementary algebra of a ninth-grade textbook. But in fact the former grew directly from the latter.

We might still say that the algebra that resulted is a "universal arithmetic," but the universality of the computations is something that Newton might not have recognized. Saying that the operation of vector addition, or the dot product of two vectors, is commutative might be seen as a generalization of the notion of commutativity for addition or multiplication of real numbers, but this generalization is far beyond what Newton describes in the passage above.

And yet there is a hint, in Newton's words, of this emphasis on operations and structures. Newton notes that algebra proceeds "in reverse." This can be construed to reflect the nature of algebraic operations, of inverses with respect to a given operation, and so on. This second step in learning algebra consists, essentially, in turning one's attention from the numbers of arithmetic to the binary operations acting on these numbers. Algebra can be thought of as the study of binary operations and their properties.

This is a big part of what Barry doesn't understand. He can substitute $x = 3$ in the expression $2x + 1$ to get 7, but he cannot solve the equation $2x + 1 = 7$ except by trial and error. He is too wrapped up in the numbers, over which his command is weak, to focus his attention on the operations he is performing. He won't add –1 to both sides of the equation, because he doesn't really understand the relationship of –1 to +1 with respect to addition and cannot see that this relationship is the same as that of 3/4 to –3/4 or that it bears similarity to the (multiplicative inverse) relationship between 3/4 and 4/3.

Merry's breakthrough consists in the fact that she sees that the operations she performs to solve $2x + 1 = 7$ are the same as those she must perform to solve $2/3x - 4 = 7/5$. But Merry can only go so far in her study of binary operations. This is because she has not taken the third step: the recognition of algebraic form.

We can get an idea of what this step means from a helpful pedagogical comment of a modern master, I. M. Gelfand. Here are his words (from a personal communication):

> Years ago, we thought of arithmetic as dealing with numbers, and algebra as dealing with letters. But we sometimes use letters as well as numbers in discussing arithmetic.
>
> A more modern view distinguishes algebra from arithmetic in another way. In algebra we let letters represent other letters, and not just numbers. That is, a student can learn the algebraic identity $A^2 - B^2 = (A + B)(A - B)$ and think of it as representing such statements as $2499 = 50^2 - 1 = (50 + 1)(50 - 1) = 51 \times 49$. This is an arithmetic statement. But if we let $A = x^6$ and $B = y^6$, the same identity can

represent the statement $x^6 - y^6 = (x^3 + y^3)(x^3 - y^3)$. Or, we could write $(\cos x)^2 - (\sin x)^2 = (\cos x + \sin x)(\cos x - \sin x)$. And so one algebraic identity spawns many others.

In mathematical terms, we can say that arithmetic is largely the study of the field of rational numbers. Algebra, however, begins with the study of the field of rational functions.

Merry does not understand this notion that the symbols of algebra can represent rational expressions. For her, each factoring problem involving the difference of two squares is a new challenge. For us, who can manipulate rational expressions easily, they are all alike.

Our traditional algebra texts are organized in just this way. Each page of drill is a set of instances of a particular identity in the field of rational expressions. When we use these drills with students, we are helping them to recognize how they are all alike and helping them to build an internal representation of their similarity. This representation is the notion of algebraic form and is a significant step in abstraction. So in one sense, Gelfand's remark, unlike Newton's, is a pedagogical note couched in mathematical terms.

Getting students to work in the field of rational expressions is an important pedagogical step in turning students' attention from the numbers of arithmetic to operations in general. It is the step that I was not able to take with Merry.

INSTRUCTIONAL INTERVENTIONS

So now I defined for myself the problems I had to solve with Barry and Merry. How could I go about trying to solve them? I had no simple answers, but only suggestions and approximations.

For Barry, I tried using a "number story" technique as a transition to algebra: there was this number *x*, and it got multiplied by 2. Then 1 got added, and the result was 7. What was the original number?

Barry knew the answer quickly if I used small numbers. So I used the same story to model the equation $2x + 1 = 247$. I reasoned, with Barry, that if 1 got added to twice the number, and the result was now 247, then twice the number, before 1 was added, must have been 246. Then, if the number was 246 after it had been doubled, it must have been 123 before it was doubled.

Barry followed this reasoning and could do a few more problems like this. But he balked when the reasoning was expressed algebraically. He dutifully wrote down the following:

$$\begin{array}{rcl} 2x + 1 & = & 247 \\ -1 & & -1 \\ 2x & = & 246 \\ \div 2 & & \div 2 \\ x & = & 123 \end{array}$$

This was just to please me. For the longest time, it had no meaning at all for Barry. Tellingly, he did not usually "get" the story solution unless I asked the questions first ("What must the number have been before we added 1?"). This situation improved with time, and Barry was eventually able to solve simple linear equations with numbers for which he could not use trial and error. An essential part of this process was played by the calculator. I would give Barry an equation like $4.1567x - 3.2564 = 7.5342$ and hold the calculator out of his reach. He had to tell me what he wanted to do with the calculator, after which I allowed him to use it. This seemed to help Barry turn his attention to the operations he was performing and away from the numbers to which the operations applied. Barry was taking step two.

Both Barry and Merry benefited from another technique involving simple rational expressions. I asked them questions of the following form:

If $2x + 3y = 3$ and if $6x + 3y = 5$, how much does $8x + 6y$ equal?

I approached this set of exercises very concretely: If 2 pens and 3 pencils cost \$3, and 6 pens and 3 pencils cost \$7, how much will 8 pens and 6 pencils cost?

They quickly found that they could not use trial and error. The problem did not call for the values of x and y. They could, conceivably, guess at the correct values for x and y, but these were fractions and not readily guessed. Just in case they could guess, I would give them problems like this:

If $x + 2y + 3z = 10$ and $2x + y = 12$, what is the value of $x + y + z$?

The students learned to manipulate these expressions without evaluating them, and so they began to think of algebraic expressions in their own right, without evaluating them. They also learned, later, to solve simultaneous equations algebraically and developed the idea of a linear combination of vectors. That is, they could explain that if $x + 2y + 3z = A$ and $2x + y = B$, then $x + y + z = 1/3(A + B)$, sometimes using exactly the representation presented in this sentence.

Barry and Merry both learned from these techniques. Merry benefited from one further technique, for which Barry was not yet ready.

This technique involved identifying the "role" of various expressions. That is, after Merry factored $4x^2 - 9y^2$ as $(2x + 3y)(2x - 3y)$, I would remind her of the identity $A^2 - B^2 = (A + B)(A - B)$ and ask her how the two problems were the same. Eventually, she was able to articulate that $2x$ "played the role of" A and $3y$ "played the role of" B. This phrase became part of her vocabulary, so that when she had trouble factoring or simplifying an expression, I would ask her to compare it with a simpler expression and ask her for the roles.

This last technique is one I find valuable for students at every level: the comprehension of algebraic form is not a process that develops simply. For more advanced students learning to tame algebraic fractions, I tell them that simplifying

$$\frac{x+1}{x+4} + \frac{x-2}{x+3}$$

is "the same as" simplifying $3/5 + 2/7$. It seems to help, and students learn from this how to decide for themselves which arithmetic pattern the problem follows. (This technique exploits the fact that both rational numbers and rational expressions have the structure of a field.)

And on a higher level still, I use the "roles" idea with students doing algebraic proofs. For example, such students often have trouble seeing that $(x + y)(p + q) = (x + y)p + (x + y)q$ is an application of the distributive law. If we state the latter as $A(B + C) = AB + AC$, asking "what plays the role of A," and so forth, things often clear up. Indeed, if students cannot answer this question, they cannot understand the proof they are reading (or even writing!).

CONCLUSION

In working with Barry and Merry, I tried to use their difficulties in learning mathematics to probe more deeply into the nature of the mathematics they were learning. I then tried to use what I found out to help them surmount their difficulties.

I have described what I learned from the process, but what did the students learn? I have already seen Merry, in her mainstream class, get further than I think she might have without my interventions. Barry didn't get as far, but I've noticed that his understanding of arithmetic has advanced, and it seems to me that it was pushed by his experience with algebra.

Neither Barry nor Merry may do any further academic work in mathematics, and neither may even need much mathematics in their careers. It seems likely to me, however, that their wrestling with the algebra as a representation of generalizations in arithmetic and then as a study of operations, and their using the abstraction of work with rational expressions, should all contribute to their understanding of abstractions in general, and this understanding is a basic and valuable cognitive tool.

REFERENCE

Whiteside, Derek Thomas, ed. *The Mathematical Papers of Isaac Newton*, vol. 5. London: Cambridge University Press, 1972.

4

Attending to Transparent Features of Opaque Representations of Natural Numbers

Rina Zazkis

Karen Gadowsky

MUCH of the discussion on different representations in mathematics education literature relates to qualitatively different representation systems, such as pictures, manipulative models, spoken language, or written symbols. The immediate associations with different representation systems bring to mind graphs, correspondence diagrams and equations (when considering functions), or written symbols and circle parts (when considering fractions). In recent discussions, representations are linked to visualization and to computer technology. Here we take a different focus. We discuss various representations of numbers and the insights they provide.

FOCUSING ON REPRESENTATIONS OF NUMBERS

Our focus is on the same symbol system—the one of numbers—and different representations that can be formed within this system. We do not address different numeration systems and, for the most part, stay within the conventional decimal representation of numbers.

Many of the definitions of various sets of numbers refer to representations. The judgment whether or not a number belongs to a given set is based on whether or not it is possible to represent it in a given form. For example, a rational number is a number that can be represented as a/b, where a is an integer and b is a nonzero integer. An even number is a number that can be represented as $2k$, where k is a whole number. A complex number is a number of the form $a + bi$, where a and b are real numbers and $i = \sqrt{-1}$. A number is a "perfect square" if it can be represented as A^2 for some natural number A. In this paper, we focus on whole numbers and the variety of ways in which to represent them.

Consider, for example, the following list :

$$(a)\, 216^2,\ (b)\, 36^3,\ (c)\, 3 \times 15552,\ (d)\, 5 \times 7 \times 31 \times 43 + 1,\ (e)\, 12 \times 3000 + 12 \times 888$$

It is not apparent that all these expressions represent the same number, 46 656. However, following Mason (1998), we say that each representation shifts our attention to different properties of the number.

Lesh, Behr, and Post (1987) describe representational systems as either transparent or opaque. A transparent representation has no more and no less meaning than the represented idea(s) or structure(s). An opaque representation emphasizes some aspects of the ideas or structures and de-emphasizes others. Opaque representations possess some properties beyond those of the ideas and structures that are embedded in them, and they do not have some properties that the underlying ideas or structures do have. In this sense, all the representations of natural numbers, including the canonical decimal representation, are opaque; however, each one has transparent features.

From representation *(a)* of 46 656, it is transparent that the number is a perfect square; representation *(b)* shows that the number is a perfect cube; from (c) we conclude that the number is a multiple of 3 and 15 552. Of course, it is possible to derive that the number is a multiple of 3 from *(a)* and *(b)* as well, but *(a)* and *(b)* do not give us a clue regarding 15 552. From *(d)*, we conclude that the number leaves a remainder of 1 in division by 5, 7, 31, and 43, a conclusion that is not apparent in representations *(a)*, *(b)*, and *(c)*. From representation *(e)*, we see that the number is a multiple of 12 and, acknowledging distributivity, that is it a multiple of 3 888.

Do students attend to these salient transparent features in different representations? Our experience with middle school students as well as with preservice elementary school teachers suggests that often this is not so. In what follows, we consider several examples that outline this experience. Students' responses were solicited using clinical interviews and written questionnaires as part of the ongoing research project on learning elementary number theory.

Attending to Factors and Multiples

Consider the number $M = 3^3 \times 5^2 \times 7$. Is M divisible by 7?

Some students gave us the "hey, isn't it obvious?" look when presented with a question like the one above. Others had to determine the value of M in order to answer the question. These students performed the multiplication using calculators, divided the result by 7, and since the result showed a whole number, they reached a positive conclusion. Students choosing this strategy understood what is required for divisibility and some of the means to reach an answer. However, they were not attending to the structure of the number given in its representation, where 7 is explicitly listed as a factor.

Students also found it difficult to determine divisibility by 7 when they were faced with the representation of M that included a multitude of factors. Some students could easily claim that the number 675×7 was divisible by 7, but they could not draw a similar conclusion from the original representation of M. This has to do with students' seeing a factor as one of two numbers they multiply together. When 7 was listed as "one of many" and not "one of two," some students simply failed to recognize it; others had to represent M as a product of two numbers, one of which was 7, in order to draw the conclusion.

An interesting phenomenon related to the representation of M in prime-factored form is not only acknowledging what is there but also noting what is not there. It was much easier for students to conclude M's divisibility by 7 than to refute M's divisibility by 11. Consider two statements:

> The number 7 is listed in prime factorization of M; therefore, M is divisible by 7.
>
> The number 11 is not listed in prime factorization of M; therefore, M is not divisible by 11.

Despite lexical similarity, reaching the second inference on the basis of a number's representation appears to be more difficult for a learner. It involves awareness of the uniqueness of prime decomposition entailed by the fundamental theorem of arithmetic.

Attending to the Structure of the Division Algorithm

> Consider the number $K = 6 \times 147 + 1$. What is the quotient and the remainder in the division of K by 6?

Students had no difficulty in finding the quotient and the remainder; however, the preferred approach to reach their conclusion was to calculate the value of K to be 883 and then perform long division to determine the quotient and the remainder.

It is transparent from the representation of number K that the quotient is 147 and the remainder is 1. In fact, this is promised by the division algorithm: For integers a and b, $b > 0$, there exist unique integers q and r such that $a = bq + r$, $0 \le r < b$. In the division of a by b, the number q is the quotient and r is the remainder. The middle school curriculum may not involve a formal introduction of the division algorithm. However, recognizing that the number 6×147 leaves no remainder in division by 6 leads to the conclusion that the next number in a sequence of natural numbers will give the remainder of 1 in division by 6. We believe that this line of reasoning is not out of reach for students familiar with what is required to perform computation. What is probably missing is the students' inclination to "think" before turning to calculations.

Recognizing Perfect and Imperfect Squares

Is 71^2 a perfect square?

Naturally. It is represented as such.

Is 71^6 a perfect square?

Not so naturally. Some students calculated the value of 71^6, extracted the square root, and reached their conclusion. Should we care about the process as long as the answer is correct? Positively. A similar strategy will not help when using a regular handheld calculator and querying whether 71^{60} is a perfect square. In these examples, it would be desirable to change the representation. A perfect square is embedded in representing the numbers mentioned above as $(71^3)^2$ or $(71^{30})^2$, respectively.

Were students familiar with computations with powers? Indeed. They had no trouble performing these computations on request. However, in this instance, many did not recognize the kind of computation required, nor did they notice what specific property they could use to make their decision.

Is 19×31 a perfect square? As earlier, we're not interested in a yes or no answer but in whether a decision can be reached "elegantly," without computation. If this is a requirement, the argument has to do with representations. In a prime factorization of a perfect square, every prime appears an even number of times. Combining this property of perfect squares with the fact that 19 and 31 are primes immediately gives a negative answer. Making the claim that a number is *not* a perfect square is more challenging, since this line of reasoning involves an awareness of the uniqueness of prime factorization.

Attending to Distributivity

Consider the number $A = 15 \times 5623 + 60$. Can you represent it as a multiple of 15?

A majority of students calculated the value of the number A and divided it by 15. A more elegant solution could represent 60 as 15×4 and apply distributivity:

$$15 \times 5623 + 60 = 15 \times 5623 + 15 \times 4 = 15 \times (5623 + 4)$$

Although there is no formal definition of what is *elegant,* there seems to be an agreement that involving extensive computation in places where it can be avoided is "not elegant." Unfortunately, regression to computation seemed to be a preferred strategy for most of our students. This could be avoided by paying attention to the transparent features of different representations.

Representation as a Pitfall

In the examples above, we attempted to make a case for paying attention to representations. In what follows, we show several examples in which doing so may not be fruitful—examples in which properties of representations were overgeneralized, applied improperly, or confused with properties of numbers.

Example 1

> Consider a number represented in a base other than ten, 121_{five}. Is it odd or even?

Some students remained trapped in the base-ten system and claimed that the number is odd, attending to its last digit. Last digits provide a helpful hint for classifying numbers as odd or even when working in the decimal representation. In this example, the salient feature of decimal representation has been overgeneralized and applied in the domain in which it does not hold. According to Kaput (1987), there is no significant attention given in the curriculum to an "important distinction between those properties of numbers that are sensitive to the representation system versus those properties that are relatively independent of the representation system" (p. 21).

Example 2

> Consider 36^3. Is it a perfect square?

Some students claimed that it was not, because it was a "perfect cube." Is there a contradiction? Obviously, in the students' minds it was one or the other. The possibility that a cube of a perfect square is a perfect square did not occur to them. It appears that in paying attention to the given representation, students did not consider the possibility of an alternative representation.

According to Harel and Kaput (1991), "notations can act as substitutes for conceptual entities, supplanting the need for them" (p. 93). We believe that examples 1 and 2 demonstrate this occurrence. In example 2, the notation A^2, for some whole number A, appeared as a substitute for a concept of the perfect square. When a desired representation was not given explicitly, the students didn't recognize the number as a perfect square.

Example 3

Reconsider the previous sections on the division algorithm and distributivity.

> What is the quotient and the remainder in division of K by 6, where $K = 6 \times 147 + 1$?
>
> Can you represent A as a multiple of 15, where $A = 15 \times 5623 + 60$?

We witnessed the following claims in the students' responses: "A is a multiple of 15 because it is written so, as 15 times something"; "K is a multiple of 6, so there is no remainder." These claims, though not representative of the majority, are not single occurrences either. Students responding in this way considered some components of representation, but not representation as a whole, and demonstrated confusion between K and $6 \times (147 + 1)$ or A and $15 \times (5623 + 60)$.

A detailed analysis of the tasks discussed above and a synopsis of students' responses to these tasks can be found in Zazkis and Campbell (1996a, 1996b), Campbell and Zazkis (1994), and Zazkis (1998).

TREATING REPRESENTATIONS OF NUMBERS AS NUMBERS

When students are asked to discuss properties of numbers represented in other than canonical decimal form, reluctance to manipulate these numbers is a common reaction. There is a tendency to "find out what the number is," that is, what its canonical decimal representation is, before the properties of the number can be mentioned. We suspect that what gets in the way of students' attending to number properties that are transparent in specific number representations is that students do not treat different representations as "numbers." The following vignette exemplifies the situation.

> Recently we presented the symbols $57 \times 1789 \times 3 + 17$ to a group of middle school students and asked them: "Is this a number?" The question presented a puzzle, possibly because of its triviality. "Yes, it is" was not a common response. Students suggested that "it can be calculated to be one" or "it equals to some number." From these suggestions, we conclude that $57 \times 1789 \times 3 + 17$ itself was not considered to be a number. If it is not a number, what is it? Suggestions included "it is an exercise" and "it is several numbers together with operations." A student explained that "7 (seven) is a number, all the rest, like $3 + 4$ or $8 - 1$ are just different ways you get this number in the answer."

We believe that the students' perception of numbers is a consequence of their prior school practices, where more emphasis had been put on calculations rather than on attention to number structure. The students' desire to calculate the canonical decimal representation can be seen both as an indication of their notion of what a number is and as a means toward empirical verifications. Returning to our first example, in order to discuss the divisibility of the number M, where $M = 3^3 \times 5^2 \times 7$, it is more convincing for many students to find out a number's canonical representation and actually perform the division with the calculator, rather than to pay attention to its factors. Such a tendency has been described by researchers as an "empirical proof scheme" (Harel and Sowder 1998).

We do not wish to emphasize the distinction between "numbers" and "numerals" attempted by the "new math" advocates. However, our intent is to help students appreciate different representations of numbers and the information they provide and not to limit the concept of a "number" to its canonical form of representation. In what follows we present several examples for classroom implementation that can combine to form a useful step in achieving this goal.

SNAPSHOTS OF ACTIVITIES FOR CLASSROOM IMPLEMENTATION

Educating learners to appreciate mathematical structure and pattern in general and structure of numbers in particular is a goal of mathematics instruction. The following activities may prove useful in considering number structure and supplement more standard activities that involve finding prime decomposition or performing operations with exponents.

Snapshot 1: Perfect Cubes and Perfect Squares

Represent each of the following numbers as a product of primes:

8 9 15 16 20 25 27 34 36 42 49 55 58 60 64 81 100 125 144

Suggest a way to classify prime factor representations from the exercise above.

We expect that among the classifications suggested by students, it will be noted that perfect squares have two identical groups of factors and perfect cubes have three identical groups of factors. If the issue is not raised by students spontaneously, the teacher could direct the students' attention to repeated prime factors. Further, students could be engaged in the following:

- Consider 36^2, 36^3, 36^4, 36^5, 36^6, 36^7. Which representations are perfect squares? Which are perfect cubes? Represent these numbers in such a way that your claim is apparent.

- Are there any perfect squares that are also perfect cubes? Explain your theory using prime-factor representation. Can you "easily" find such numbers? What would be your strategy?
- How could all perfect squares and perfect cubes between 100 and 1000 be found without using a calculator and without engaging in extensive paper-and-pencil calculations? Record your initial ideas in your journal. Discuss your ideas with others and revise your initial strategy if necessary.

Once prime-factor representation is discussed and used as a strategy, many students seem able to embark on this thinking pathway. For students who cannot overcome their tendency to check all their inferences with a calculator, we suggest going beyond the calculator abilities:

- Consider 36^{200}, 36^{300}, 36^{400}, 36^{500}, 36^{600}, 36^{700}.

Which representations are perfect squares? Which are perfect cubes? Represent these numbers such that your claim is apparent.

- Consider 2^{100}, 3^{100}, 2^{99}, 3^{99}

Are these numbers odd or even? Could they be perfect squares or perfect cubes? Justify your responses.

Snapshot 2: Divisibility by 13

Consider the following numbers:

$$13^{50},\ 50^{13},\ 1\,000\,000\,000,\ 123\,456 \times 13 + 3,\ 39 + 654\,321 \times 13,$$
$$36 \times 7654 + 3 \times 4567,\ 2^4 \times 3^5 \times 5^6,\ 2^4 \times 13^5 \times 5^6$$

For which of these numbers is it possible to decide whether they are divisible by 13 without performing calculations? Justify your decision. Make up another five numbers in which their divisibility or indivisibility by 13 can be derived by attending to their representation. Trade these numbers with a partner and check whether you agree about their divisibility.

Integrating activities similar to those above in the classroom fosters students' attention to transparent features of different representations of numbers.

CONCLUSION

"Capitalizing on the strengths of a given representation is an important component of understanding mathematical ideas" (Lesh, Behr, and Post 1987, p. 56). Helping students recognize patterns in different representations and situations in which paying attention to these patterns is appropriate is our goal. We suggest that a discussion of the properties of numbers in which we consider their noncanonical representations is a valuable pedagogical activity. It proves to be especially stimulating when numbers presented to students were beyond the computational abilities of a handheld calculator. For many students, the inability to check their inference with a calculator presented inconvenience and challenge.

Careful pedagogical choice of activities can help students identify transparent features of different representations of numbers. Attending to these features may help students achieve a better understanding of the multiplicative structure of natural numbers and appreciate these representations as "numbers" with the respect they deserve.

REFERENCES

Campbell, Stephen, and Rina Zazkis. "The Distributive Flaws: Latent Conceptual Gaps in Preservice Teachers' Understanding of the Property Relating Multiplication to Addition." In *Proceedings of the Sixteenth Annual Meeting of the North American Chapter of the International Group for the Psychology of Mathematics Education,* edited by David Kirshner, pp. 268–74. Baton Rouge, La., 1994.

Harel, Guershon, and James Kaput. "The Role of Conceptual Entities and Their Symbols in Building Advanced Mathematical Concepts." In *Advanced Mathematical Thinking,* edited by David Tall, pp. 82–94. Boston: Kluwer Academic Publishers, 1991.

Harel, Guershon, and Larry Sowder. "Students' Proof Schemes: Results from Exploratory Studies." In *Research in Collegiate Mathematics Education,* Vol. 3, edited by Alan H. Schoenfeld, James Kaput, and Ed Dubinsky, pp. 234–83. Providence, R.I.: American Mathematical Society, 1998.

Kaput, James. "Representation Systems and Mathematics." In *Problems of Representation in the Teaching and Learning of Mathematics,* edited by Claude Janvier, pp. 19–28. Hillsdale, N.J.: Lawrence Erlbaum Associates, 1987.

Lesh, Richard, Merlyn Behr, and Thomas Post. "Rational Number Relations and Proportions." In *Problems of Representation in the Teaching and Learning of Mathematics,* edited by Claude Janvier, pp. 41–58. Hillsdale, N.J.: Lawrence Erlbaum Associates, 1987.

Mason, John. "Enabling Teachers to Be Real Teachers: Necessary Levels of Awareness and Structure of Attention." *Journal of Mathematics Teacher Education* 1 (1998): 243–67.

Zazkis, Rina. "Odds and Ends of Odds and Evens: An Inquiry into Students' Understanding of Even and Odd Numbers." *Educational Studies in Mathematics* 36 (June 1998): 73–89.

Zazkis, Rina, and Stephen Campbell. "Divisibility and Multiplicative Structure of Natural Numbers: Preservice Teachers' Understanding." *Journal for Research in Mathematics Education* 27 (November 1996a): 540–63.

———. "Prime Decomposition: Understanding Uniqueness." *Journal of Mathematical Behavior* 15 (June 1996b): 207–18.

5

The Influence of Language on Mathematical Representations

Irene T. Miura

In the classroom, there are two general types of representations that affect children's understanding of, and solutions to, mathematics problems: (1) instructional representations (e.g., definitions, examples, and models) that are used by teachers to impart knowledge to students and (2) cognitive representations that are constructed by the students themselves as they try to make sense of a mathematical concept or attempt to find a solution to a problem. The first representations are external to the student (i.e., shared means of communication between the instructor and the learner), whereas the second are internal to the student (and may be unshared with others).

Both types of representations are influenced by cultural factors, including the characteristics of the language used in the mathematics domain. Some languages may provide better support than others for particular instructional representations, making the underlying concepts easier to understand. Connections between the mathematical idea and the instructional representation or between the mathematical notation and the representation may be easier to discern because of certain characteristics of the verbal language. Cognitive representations may also be directly influenced by characteristics of specific languages.

Mathematical activities are set in cultural contexts with their own tools for thinking and learning, one important tool being the language of mathematics (Kaput 1991; Rogoff 1990; Steffe, Cobb, and von Glasersfeld 1988). Culturally developed tools or symbol systems restructure mental activities without altering basic abilities such as memory or logical reasoning (Nunes 1992). Nunes (1997) describes this facilitation as *mediated action;* certain tasks are made easier because of tools developed to perform the task, and these tools differ by cultural group. These culturally developed tools may also influence mathematical representations used in thinking about mathematics concepts and in solving mathematical problems. An example is numeration systems (like the English-language base-ten system) that allow humans to go beyond their natural memory capacities to count large numbers of objects.

How children think about numbers and other mathematical concepts is influenced by the characteristics of mathematical terms peculiar to their own language. Teachers' choices of instructional representations are also affected by language. Three examples of the influence of language characteristics on mathematics understanding and performance are provided: (1) the number-naming (counting) system, (2) terms for fractions, and (3) the use of numeral classifiers.

THE INFLUENCE OF NUMBER-NAMING SYSTEMS ON THE COGNITIVE REPRESENTATION OF NUMBER

The counting system in languages that are rooted in ancient Chinese (among them, Chinese, Japanese, and Korean) are organized so that they are congruent with the traditional base-ten numeration system. In this system, the value of a given digit in a multidigit numeral depends on the face value of the digit (0 through 9) and on its position in the numeral, with the value of each position increasing by powers of 10 from right to left. The spoken numerals in western languages (e.g., eleven, twelve, and twenty in English) may lack the elements of tens and ones that are contained in them. Also, the order of spoken and written numerals may not agree (e.g., fourteen for 14 in English). In Chinese, Japanese, and Korean, 11 is read (and spoken) as ten-one, 12 as ten-two, 14 as ten-four, and 20 as two-ten(s). Fifteen and 50, which are phonologically similar in English, are differentiated; 15 is spoken as ten-five and 50 as five-ten(s). Plurals are tacitly understood. Thus, the spoken numeral corresponds exactly to the implied quantity represented in symbolic form (table 5.1); the syntax and semantics are, as Cobb (1995) suggests, reflexively related.

A task constructing numbers using base-ten blocks (unit blocks and tens blocks, bars that have ten segments marked on them) was designed to explore children's cognitive representation of number (Miura 1987). Children in the first half of first grade, with no prior experience or instruction in using base-ten blocks, were shown a numeral on a card and asked to construct the number using the blocks. French, Swedish, and U.S. first graders showed an initial preference for representing the numeral using a one-to-one collection (e.g., 42 unit blocks for the numeral 42). Chinese, Japanese, and Korean first graders, however, represented the same numeral with a canonical base-ten representation (4 tens blocks and 2 unit blocks). For five numerals (11, 13, 28, 30, and 42), the Asian-language speakers used more canonical base-ten representations than the non-Asian-language speakers,

TABLE 5.1
Number Names in Four Languages

Number	English	Chinese	Japanese	Korean
1	one	yi	ichi	il
2	two	er	ni	ee
3	three	san	san	sam
4	four	si	shi	sah
5	five	wu	go	oh
6	six	liu	roku	yook
7	seven	qi	shichi	chil
8	eight	ba	hachi	pal
9	nine	jiu	kyu	goo
10	ten	shi	juu	shib
11	eleven	shi-yi	juu-ichi	shib-il
12	twelve	shi-er	juu-ni	shib-ee
13	thirteen	shi-san	juu-san	shib-sam
14	fourteen	shi-si	juu-shi	shib-sah
15	fifteen	shi-wu	juu-go	shib-oh
16	sixteen	shi-liu	juu-roku	shib-yook
17	seventeen	shi-qi	juu-shichi	shib-chil
18	eighteen	shi-ba	juu-hachi	shib-pal
19	nineteen	shi-jiu	juu-kyu	shib-goo
20	twenty	er-shi	ni-juu	ee-shib
21	twenty-one	er-shi-yi	ni-juu-ichi	ee-shib-il
22	twenty-two	er-shi-er	ni-juu-ni	ee-shib-ee
30	thirty	san-shi	san-juu	sam-shib
40	forty	si-shi	shi-juu	sah-shib
50	fifty	wu-shi	go-juu	oh-shib
60	sixty	liu-shi	roku-juu	yook-shib
70	seventy	qi-shi	shichi-juu	chil-shib
80	eighty	ba-shi	hachi-juu	pal-shib
90	ninety	jiu-shi	kyu-juu	goo-shib

and, overall, the Asian-language speakers used fewer one-to-one collection representations than the others (Miura et al. 1988; Miura et al. 1993). When asked if they could show the number in a different way (using the blocks), the Asian-language speakers were better able than non-Asian-language speakers to make two different constructions for each number, suggesting better understanding and greater flexibility in dealing with number quantities.

Effects on Place-Value Understanding

A test of children's place-value understanding found a positive correlation between the use of canonical constructions to represent numbers and the understanding of place-value concepts (the meaning assigned to individual units in a multidigit numeral). On a set of five place-value tasks, Japanese and Korean first graders showed significantly greater understanding of place-value concepts than French, Swedish, and U.S. children (Miura et al. 1993). All the Japanese and Korean first graders completed at least one problem correctly; 42 percent of Japanese children and 54 percent of the Korean first graders were able to solve all five problems correctly. By contrast, the French, Swedish, and U.S. children in the study performed significantly less well; 50 percent of the U.S. first graders could not solve any of the problems correctly.

Effects on Counting Performance

In Miller and Stigler's (1987) study of counting performance, Chinese-speaking children showed the same pattern in the development of counting skills as U.S. preschoolers did. However, the Chinese children made significantly fewer errors in number naming than U.S. children did. The activity of generating number labels poses little problem for Chinese children, and in this aspect of counting, they surpass their U.S. counterparts. Because counting in the early grades is also a tool for problem solving, it is a skill that can directly affect mathematics performance. In addition, as children learn to count and to make sense of the counting numbers, the regularity of the Asian number-naming system cannot help but foster a deep understanding of what the numbers mean.

In a study comparing six-year-old Taiwanese and English children's understanding of additive composition and counting, Lines and Bryant (in Nunes and Bryant 1996) reported that Chinese speakers showed a distinct advantage over English speakers in counting and in combining units of different sizes (in this instance 1, 5, and 10); this was particularly true for additive composition involving decades. The authors concluded that the regularity and transparency of the Chinese counting system appeared to facilitate children's learning significantly. This influence seems to result from the linguistic cues provided by the number-naming system and from a general understanding of additive composition that is supported by the system.

Effects on Addition and Subtraction Performance

When counting in Japanese (also in Chinese and Korean), one counts from 1 to 10, and then the 1 to 9 is reinforced in the numerals 11 through 19. As a result, mathematics tasks around tens are easier because the child deals with 1 to 9 only. In addition, children do not have to learn novel decade-number names, a requirement that often hinders U.S. children's counting.

In the specific area of two-digit addition and subtraction with regrouping (borrowing and carrying) that is taught at the second-grade level in both Japan and the United States, the Japanese number-naming system may have an important effect. In the addition algorithm, 59 + 8 = ?, the numbers in the ones column total 17. English speakers must ask themselves if they can regroup or trade to make a ten. If the answer is yes, this results in a regrouped 1 ten and 7 ones. The 7 is written in the ones column, and the 1 (which represents 1 ten) is added to the 5 in the tens column. In Japanese, the sum of the numbers in the ones column (17) is spoken as ten-seven. Therefore, it is readily apparent that a ten has been formed, and 1 ten is carried to the tens column, eliminating the intermediate assessment step and keeping the meaning of the individual digits in the solution intact.

Consider the following subtraction problem, 67 – 59 = ?. There are at least two possible solutions for this problem after determining that there are not enough units in the ones column to solve the problem without regrouping. First, 1 ten from the tens column can be traded or regrouped for 10 ones to make the number in the ones column a 17. Then, the problem is solved by subtracting 9 from 17 to equal 8. If the regrouped 17 is spoken as ten-seven, a second solution becomes apparent: 9 can be subtracted from the 10, leaving a remainder of 1, which is then added to the 7 to equal 8 in the ones column.

U.S. children are taught to solve the subtraction algorithm using the trading or regrouping solution (the terms vary by textbook publisher). The second solution can be used easily by East Asian children because their number language treats the teen numbers (11–19) as 10 + the single digit; thus, the two numbers (10 and the single digit) can be dealt with separately in the solution process. English-speaking children must learn the addition and subtraction facts to 18, whereas Asian-language-speaking children do not have to master combinations beyond 10.

An examination of textbooks shows that multidigit addition and subtraction with regrouping are introduced earlier in East Asian countries than in the United States (Fuson, Stigler, and Bartsch 1988). The authors suggest that less time may have to be spent on developing a foundation of place-value understanding required to perform such mathematics because the base-ten structure of the number-naming systems provides the necessary support.

The Influence of Fraction Names on the Understanding of Geometric Part-Whole Representations

Language characteristics may also influence instructional representations used by teachers to convey knowledge to their students. The concept of fractions as parts of a region and their connection to representations using geometric shapes may be easier to understand when spoken in East Asian languages. In these languages, the concept of fractional parts is embedded in the mathematics terms used for fractions. For example, in Japanese, one-third is spoken as *san bun no ichi,* which is literally translated as "of three parts, one." Thus, unlike the English word *third,* the Japanese term, *san bun* (three parts), directly supports the concept of the whole divided into three parts. The term focuses on the number of parts in a whole and connects the fraction symbol to a mathematical meaning of fractions. In teaching numerical fractions, U.S. textbooks include drawings of geometric figures divided into parts, with one or more of the parts distinguished from the rest by being shaded. Along with the visual representation, the text may explain that this fraction means

$$\frac{1 \text{ of the}}{3 \text{ equal parts.}}$$

Children enter school with an informal or intuitive understanding of rational numbers (Hunting and Sharpley 1988; Kieren 1988; Mack 1992), and language characteristics may play a role in the development of these early conceptualizations. To determine if the Asian (in this instance, Korean) vocabulary of fractions might influence the meaning children ascribe to numerical fractions, and if this would result in these children being able to associate numerical fractions with their corresponding pictorial representations prior to formal instruction, groups of first and second graders in Croatia, Korea, and the United States participated in a study at three separate times: the middle of grade 1, the end of grade 1, and the beginning of grade 2 (Miura et al. 1999). An examination of textbooks showed that the children had not had formal instruction in fraction concepts prior to the testing. The children were given seven written fractions (1/3, 2/3, 2/4, 3/4, 2/5, 3/5, and 4/5), with one repeated, for a total of eight items; each numerical fraction was followed by four geometric figures (circles, squares, or rectangles) with varying portions shaded. The fractions were read aloud by the teacher, and the children were asked to draw a circle around the picture in the row that showed the fraction.

There was a developmental difference in the children's performance, with Korean children better able at all levels to associate complex numerical fractions with their pictorial representations. The difference was only marginally significant at the middle of grade 1, but at the end of grade 1 and the beginning of grade 2, the Korean children's performance was significantly better than that of the children in Croatia and the United States. The results suggest that the Korean language may influence the meaning these children give to numerical fractions and that this enables children to associate numerical fractions to their corresponding pictorial representations prior to formal instruction. The terms match symbolic representation (e.g., 1/3) to pictorial representation (geometric region) and make connections among fraction symbols, verbal terms, and informal knowledge. An examination of mathematics textbooks in the three countries indicates that fraction concepts are introduced earlier in Korea (in grade 2) than in Croatia or the United States.

Whether this early understanding of fractions as parts of a region affects subsequent knowledge of rational number concepts is an area that needs further exploration.

THE INFLUENCE OF NUMERAL CLASSIFIERS IN SOLVING ARITHMETIC WORD PROBLEMS

As students read or listen to arithmetic word problems, they must construct cognitive representations of aspects of the problem in order to be able to find the answer to the question posed. *Numeral classifiers,* a linguistic convention used when enumerating (or counting) objects in Japanese and other Asian languages, may influence this representational process. This linguistic form (numeral classifier) is a special morpheme that indicates certain semantic features of whatever is being counted (Sanches 1977). For example, when counting five pencils in Japanese, one cannot use the cardinal number alone. One must say, five (long, thin thing) pencil, or *go-hon,* where *go* is 5 and *hon* is the numeral classifier for long, thin objects. Which classifier is used depends on the perceptual characteristics of the objects or on distinctions between human and other animal forms (Naganuma and Mori 1962). This is similar to the situation in English when counting a quantity of material or mass (e.g., water, paper, or cattle). In English, it is necessary to say one drop of water, two sheets of paper, or three head of cattle rather than one water, two papers, and three cattles.

Numeral classifiers are a grammatical convention. The numeral classifier serves as a sorting mechanism denoting cognitive categories (Denny 1986). With respect to mathematics, the numeral classifier may add a coherence to the items being enumerated; it may act to integrate the items into a set rather than treating them as individual items. Numeral classifiers also may serve as placeholders in that they are used anaphorically to denote referents that have already been introduced in the text (Downing 1986).

The word problem in English

Joe has 6 marbles.

He has 2 more than Tom.

How many does Tom have?

would be translated into Japanese as

Joe has 6 *(ko)* marbles.

2 *(ko)* more than Tom.

How many *(ko)* does Tom have?

In the Japanese-language version, the numeral classifier, *ko,* must be used for counting marbles, which are in the category of small, round objects. The problem is understood as "Joe has 6 (small, round objects) marbles. 2 (small, round objects) more than Tom. How many (small, round objects) does Tom have?"

The experienced English-speaking problem solver has learned that the 2 in the second sentence refers to marbles owned by Joe and their relationship to the number of marbles owned by Tom. The problem solver must also understand that the question posed refers to marbles and not to something else in Tom's possession. There is no confusion in the Japanese version. In Japanese, nouns, but not the corresponding numeral classifiers, may be omitted from numerical phrases once the referent has been established. Thus, the use of the numeral classifier may make problems more concrete; numbers in isolation (as in the phrase, "2 more than Tom") are not an abstract quantity. The numeral classifier acts as a concept signifier and may also serve to engage children in a stronger cognitive representation of what the story problem is asking (e.g., by removing the ambiguity from the final question).

SUMMARY

Culturally specific characteristics of mathematical languages serve to facilitate mathematics activities in a variety of ways. The Asian number-naming system, which is regular and transparent, may help children to develop a deep understanding of number concepts, including the understanding of place-value concepts. The number-naming system also provides linguistic support to facilitate abstract counting and addition and subtraction with regrouping. Because the spoken multidigit numerals in these Asian languages reflect the base-ten numeration system, numbers (especially those from 11 to 19, which are particularly difficult in non-Asian languages) are readily generated and their component parts easily understood. Children, when solving problems requiring regrouping, can deal with ten and the additional units separately in their computations.

Specific mathematics terms, such as the spoken words for fractions, may provide support for making connections between symbolic and pictorial representations. Asian-language speakers exhibit an intuitive (or socially acquired) knowledge of numerical fraction concepts prior to school instruction that suggests an understanding of fractions as parts of regions depicted by geometric representations.

An additional linguistic characteristic of Asian languages and their possible influence on mathematics understanding and performance is the use of numeral classifiers, which are perceptual or categorical in nature and are used when enumerating objects. These numeral classifiers, as part of arithmetic story problems, may diminish ambiguity by making the referent clear, add coherence to items and assign them to a set, and provide a stronger visual representation of what a problem is asking.

These culturally determined characteristics of mathematical languages may influence the choice, application, and effectiveness of instructional representations used by teachers in the educational process. They may also affect the cognitive representations constructed by students as they attempt to understand mathematical concepts and engage in problem-solving activity and, by doing so, mediate the development of conceptual knowledge in this domain.

REFERENCES

Cobb, Paul. "Cultural Tools and Mathematical Learning: A Case Study." *Journal for Research in Mathematics Education* 26 (July 1995): 362–85.

Denny, J. Peter. "The Semantic Role of Noun Classifiers." In *Noun Classes and Categorization: Proceedings of a Symposium on Categorization and Noun Classification,* edited by Colette Craig, pp. 295–308. Amsterdam: Benjamins, 1986.

Downing, Pamela. "Anaphoric Use of Classifiers in Japanese." In *Noun Classes and Categorization: Proceedings of a Symposium on Categorization and Noun Classification,* edited by Colette Craig, pp. 345–75. Amsterdam: Benjamins, 1986.

Fuson, Karen C., James W. Stigler, and Karen Bartsch. "Grade Placement of Addition and Subtraction Topics in Japan, Mainland China, the Soviet Union, Taiwan, and the United States." *Journal for Research in Mathematics Education* 19 (November 1988): 449–56.

Hunting, Robert P., and Christopher F. Sharpley. "Fraction Knowledge in Preschool Children." *Journal for Research in Mathematics Education* 19 (March 1988): 175–80.

Kaput, James J. "Notation and Representation." In *Constructivism in Mathematics Education,* edited by Ernst von Glasersfeld, pp. 53–74. Dordrecht, Netherlands: Kluwer, 1991.

Kieren, Thomas E. "Personal Knowledge of Rational Numbers: Its Intuitive and Formal Development." In *Number Concepts and Operations in the Middle Grades,* edited by James Hiebert and Merlyn Behr, pp. 162–81. Hillsdale, N.J.: Lawrence Erlbaum Associates, and Reston, Va.: National Council of Teachers of Mathematics, 1988.

Mack, Nancy K. "Learning Rational Numbers with Understanding: The Case of Informal Knowledge." In *Rational Numbers: An Integration of Research,* edited by Thomas P. Carpenter, Elizabeth Fennema, and Thomas A. Romberg, pp. 85–106. Hillsdale, N.J.: Lawrence Erlbaum Associates, 1992.

Miller, Kevin F., and James W. Stigler. "Counting in Chinese: Cultural Variation in a Cognitive Skill." *Cognitive Development* 2 (1987): 279–305.

Miura, Irene T. "Mathematics Achievement as a Function of Language." *Journal of Educational Psychology* 79 (1987): 79–82.

Miura, Irene T., Chungsoon C. Kim, Chih-Mei Chang, and Yukari Okamoto. "Effects of Language Characteristics on Children's Cognitive Representation of Number: Cross-National Comparisons." *Child Development* 59 (1988): 1445–50.

Miura, Irene T., Yukari Okamoto, Chungsoon C. Kim, Marcia Steere, and Michel Fayol. "First Graders' Cognitive Representation of Number and Understanding of Place Value: Cross-National Comparisons—France, Japan, Korea, Sweden, and the United States." *Journal of Educational Psychology* 85 (1993): 24–30.

Miura, Irene T., Yukari Okamoto, Vesna Vlahovic-Stetic, Chungsoon C. Kim, and Jong H. Han. "Language Supports for Children's Understanding of Numerical Fractions: Cross-National Comparisons." *Journal of Experimental Child Psychology* 74 (1999): 356–65.

Naganuma, Naoe, and Kiyoshi Mori. *Practical Japanese.* Tokyo, Japan: Tokyo School of the Japanese Language, 1962.

Nunes, Terezinha. "Cognitive Invariants and Cultural Variation in Mathematical Concepts." *International Journal of Behavioral Development* 15 (1992): 433–53.

———. "Systems of Signs and Mathematical Reasoning." In *Learning and Teaching Mathematics: An International Perspective,* edited by Terezinha Nunes and Peter Bryant, pp. 28–44. East Sussex, U.K.: Psychology Press, 1997.

Nunes, Terezinha, and Peter Bryant. *Children Doing Mathematics.* Oxford, U.K.: Blackwell Publishers, 1996.

Rogoff, Barbara. *Apprenticeship in Thinking: Cognitive Development in Social Context.* New York: Oxford University Press, 1990.

Sanches, Mary. "Language Acquisition and Language Change: Japanese Numeral Classifiers." In *Sociocultural Dimensions of Language Change,* edited by Ben G. Blount and Mary Sanches, pp. 51–62. New York: Academic Press, 1977.

Steffe, Leslie P., Paul Cobb, and Ernst von Glasersfeld. *Construction of Arithmetical Meanings and Strategies.* New York: Springer-Verlag, 1988.

6

The Role of Tools in Supporting Students' Development of Measuring Conceptions

Michelle Stephan

Paul Cobb

Koeno Gravemeijer

Beth Estes

CLASSROOM teachers have been encouraged to use a variety of tools to aid students' mathematical learning. For instance, the NCTM *Professional Standards for Teaching Mathematics* (1991) emphasizes that "teachers must value and encourage the use of a variety of tools" (p. 52) in order to promote discourse that focuses more on mathematical ideas than on observable calculations and methods. This article describes the results of a classroom teaching experiment in which a number of tools were designed to support first graders' development of increasingly sophisticated measuring conceptions, and the following two main points will be discussed in detail.

First, for us, there has been an inherent tension between building on students' personally constructed mathematical tools and introducing adult-designed devices. In this paper, we will describe our way of dealing with this tension. We will illustrate that each of the tools used in the first-grade class-

The members of the project team included Paul Cobb, Beth Estes, Koeno Gravemeijer, Kay McClain, Beth Petty, Michelle Stephan, and Erna Yackel. The investigation reported in this paper was supported by the National Science Foundation under grant no. REC 9814898 and by the Office of Educational Research and Improvement under grant no. R305A60007. The opinions expressed do not necessarily reflect the views of either the Foundation or OERI.

room teaching experiment was either created by the students with the skillful guidance of the teacher or was introduced by the teacher as a natural solution to a dilemma with which the students were grappling. Introducing tools in each of these ways is our attempt at dealing with the tension between students' creativity and the adult designers' intent.

A second point we will make is that each tool used in the experiment was considered a part of a coherent sequence of instructional activities. Often, a tool is designed as a stand-alone device to teach a mathematical concept. Instead, our approach was to design a series of tools, each of which built on the mathematical learning that students constructed as they reasoned with previous tools. This paper is organized as follows. First, we describe the first-grade classroom that is the subject of this paper. Second, we present examples from the classroom to illustrate the two aspects of tool use discussed above.

THE CLASSROOM

The first-grade classroom that was the subject of this study was one of four first-grade classrooms at a private school in Nashville, Tennessee. The class consisted of sixteen children, seven girls and nine boys. The majority of the students were from middle-class backgrounds. The classroom teaching experiment took place over a four-month period from February to May 1996. Just prior to the teaching experiment, the teacher and the students had been engaged in instruction on single-digit addition and subtraction.

The purpose of the classroom teaching experiment was to design and improve two closely related instructional sequences. The first sequence was designed to support the students' development of increasingly sophisticated measuring conceptions. This sequence was designed in such a way as to serve as a basis for a second instructional sequence that supported the students' thinking strategies for two-digit addition and subtraction. Our goal was that the activity of measuring would not merely be a matter of iterating the units on a ruler or some other measurement tool and verbalizing the number obtained when the measurement is iterated for the last time. We wanted the number that results from the last iteration to signify not simply the last iteration itself but rather the result of the accumulation of the distances iterated (cf. Thompson and Thompson [1996]). For example, if students were measuring by pacing heel to toe, we hoped that the number words they said as they paced would indicate the distance paced thus far rather than the single pace that they made as they said a number word.

Another goal was to help the students think about spatial distances flexibly. For instance, our intent was that students would be able to interpret their measuring activity as not only a space measuring 25 feet but also five distances of 5 feet or two distances of 10 feet and a distance of 5 feet, as the need arose.

Because of space constraints, the examples that we present focus strictly on the tools created and used during the measurement sequence. The examples illustrate both the importance we place on sequencing tools so that they build on the students' prior knowledge and how the teacher in each example continually built on the students' contributions as new tools were introduced.

CLASSROOM EPISODES

Example 1: The King's Foot

The instructional activities used in the teaching experiment were typically posed in the context of an ongoing narrative. The teacher engaged students in a story in which the characters encountered various problems that the students were asked to solve. She began the first narrative by describing that a king wanted to measure items in his kingdom using only his foot. The teacher asked the class how the king could use his foot to measure. The students made several proposals, and the teacher capitalized on a student's suggestion to count each foot as he placed one in front of the other. The teacher asked the students to pretend they were king and to find the length of different items around the classroom. Although the teacher had pacing in mind as a starting point for measuring prior to the whole-class discussion, she engaged students in the narrative so that measuring would be experienced by the students as a solution to a problem.

Initially, students measured the length of items, such as a rug, in two different ways. Some students placed one foot at the beginning of the rug and counted "one" with the placement of their second foot (see fig. 6.1a). Other students placed their foot at the beginning of the same rug and counted it as "one" (see fig. 6.1b). As we talked with the teacher, we conjectured that students who were counting their paces in the first manner (fig. 6.1a) were not thinking about their paces as space-covering units. In other words, they did not regard the first foot they had placed at the beginning of the rug as something that was covering part of the rug. For them, measuring was not about covering the space defined by the carpet. Rather, the goal of measuring seemed to be to count, in some manner, the number of paces it took to reach the end of the carpet. However, it would be a mistake to claim that students

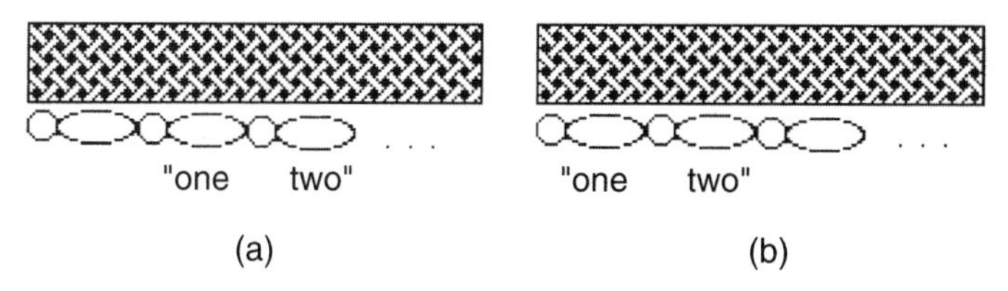

Fig. 6.1. Two methods of counting as students paced the length of a rug

who measured the rug in the second manner (fig. 6.1b) *were* thinking about covering space with each pace.

The teacher capitalized on the students' differing solutions by asking them to compare and contrast these two methods of measuring during the subsequent whole-class discussion. In doing so, she hoped that measuring as "covering space" might become an explicit topic of conversation. The intent of these whole-class discussions was not to make students count their paces in "the right way." Rather, the teacher believed that if the students participated in a discussion about these two ways of counting their paces, they would have an opportunity to reorganize their understanding about what it means to measure.

Negotiating the meaning of measuring

The teacher began the subsequent whole-class discussion by asking two students, whom she had observed counting their paces in the alternative ways described above, to pace the length of the rug for the rest of the students. In the excerpt below, both the method and purpose of pacing were negotiated by the teacher and the students. The teacher asked Sandra and Alice to show the class how each of them would measure the rug.

T: I was also really watching how a couple of you were measuring. Who wants to show us how you'd start off measuring, how you'd think about it?

Sandra: Well, I started right here *[places the heel of her first foot at the beginning of the rug]* and went 1 *[starts counting with the placement of her second foot as in fig. 6.1a]* 2, 3, 4, 5, 6, 7, 8.

T: Were people looking at how she did it? Did you see how she started? Who thinks they started a different way? Or did everybody start like Sandra did? Alice, did you start a different way or the way she did it?

Alice: Well, when I started, I counted right here, *[places the heel of her first foot at the beginning of the rug and counts it as "one" as in fig. 6.1b]*, 1, 2, 3.

T: Why is that different from what she did?

Alice: She put her foot right here *[places it next to the rug]* and went 1 *[counts "one" as she places her second foot]*, 2, 3, 4, 5.

T: How many people understand that Alice says that what she did and what Sandra did were different? How many people think they understand? Do you think you agree they've got different ways?

In response to the teacher's last questions, many students indicated that they could not see a difference between the two methods. The problem in understanding, as well as communicating, the difference between the two methods lies in the difficulty in distinguishing which paces were (were not) being counted. When a student counted a pace as "one," she must lift that foot and place it ahead of her second foot to count it as "three." As soon as the "one" foot is lifted, the record of the first pace disappears. Therefore, after the students had paced three or four paces, it was difficult for (1) the student who was pacing to communicate what "one" referred to and (2) other students to see what the demonstrator meant by the first pace.

The emergence of the first tool

In order to support a conceptual discussion of these two different methods, the teacher placed a piece of masking tape at the beginning and end of each pace. Once a record of four or five paces was made, the students who counted their paces Alice's way began to argue that Sandra's method of counting would lead to a smaller result because she did not count the first foot. In the excerpt below, Melanie differentiated between the two ways of counting paces while other students justified their particular method.

Melanie: Sandra didn't count this one *[puts foot in first taped space]*, she just put it down and then she started counting 1, 2. She didn't count this one, though *[points to the space between the first two pieces of tape]*.

T: So she would count 1, 2 *[refers to the first three spaces, since the first space is not being counted by Sandra]*. How would Alice count those *[points to the first three taped spaces]*?

Melanie: Alice counted them 1, 2, 3.

T: So for Alice there's 1, 2, 3 there and for Sandra there's 1, 2.

Melanie: Because Alice counted this one *[points to the first taped space]* and Sandra didn't, but if Sandra would have counted it, Alice would have counted three and Sandra would have too. But Sandra didn't count this one so Sandra has one less than her.

T: What do you think about those two different ways, Sandra, Alice, or anybody else? Does it matter? Or can we do it either way? Hilary?

Hilary: You can do it Alice's way or you can do it Sandra's way.

T: And it won't make any difference?

Hilary: Yeah, well, they're different. But it won't make any difference because they're still measuring but just a different way and they're still using their feet. Sandra's leaving the first one out and starting with the second one, but Alice does the second one and Sandra's just calling it the first.

⋮

Phil: She's 15 *[refers to the total number of feet Sandra counted when she paced]*. Alice went to the end of the carpet *[he means the beginning of the carpet]*. Sandra started after the carpet. Hers is lesser 'cause there's lesser more carpet. Alice started here and there's more carpet. It's the same way but she's ending up with a lesser number than everybody else.

Alex: She's missing one right there. She's missing this one right here *[points to the first taped space]*. She's going one but this should be one *[the first taped space]* 'cause you're missing a foot so it would be shorter.

T: So he thinks that's really important. What do other people think?

Alex: Since you leave a spot, it's gonna be a little bit less carpet.

This episode highlights the important role that the record created by using the masking tape played in the discussion. Before the record was made, students had difficulty distinguishing between the two methods of measuring the length of the rug. However, once the record of paces was available, the students could point to particular spaces that were or were not being counted. In this way, the record supported a significant change in the discussion.

In the first episode we presented, the students were primarily concerned with how paces were counted. Now, as students referred to the record of paces, they began talking about the amount of space that was being measured. Phil and Alex both gave explanations in terms of an amount of carpet (e.g., "shorter") rather than the number of paces counted. As Alex noted in the last line of the excerpt, "Since you leave a spot, it's gonna be a little bit less carpet." Thus, the record of paces became a means of organizing space for students. With the teacher's guidance, the symbolic record facilitated discussions in which measuring as covering space was a topic of conversation.

Note that the students' informal notions of counting paces was an appropriate starting point for the measurement sequence. The dilemma of the king appeared to be a story that students could imagine very easily, and pacing as an initial basis of measuring was a realistic solution to the king's problems (see also Lubinski and Thiessen [1996]). The teacher built on the students' solutions and introduced the masking tape as a record of their pacing activity. In doing so, the teacher introduced the first tool (record of pacing) in a natural way as a solution to the problem the students were having in communicating their methods of measuring. The record was consistent with the students' mathematical activity at that point and was readily accepted by the students as a helpful way of describing their varying interpretations.

As the instructional activities involving pacing continued, whole-class discussions began to focus on how students counted their last few paces when measuring an item. Significantly, the students did not extend their last foot beyond the endpoint of the item they were measuring. If a student was pacing along the edge of the carpet, for instance, and her last pace extended past the end of the rug, she typically did not place the last foot down but instead estimated "14 and a little space." Other students actually counted "14" and turned their foot sideways (saying either "15" or "14 1/2") so that it fit exactly. This suggests that measuring was an activity that involved covering the physical extension of an object exactly without any part of a foot extending beyond the object (see Stephan [1998]). In the next example, as students reasoned with a new tool, the interpretation of measuring as covering the object *exactly* was explicitly challenged.

Example 2: Measuring with a Footstrip

Five days after the introduction of the king's foot scenario, the teacher explained that the king could not be everywhere at one time. Thus, the students' job was to think about ways that the king could use his foot to measure items in his kingdom without doing all the measuring himself. The teacher explained that one of the king's advisors had suggested that the king trace his feet on paper. One student added that he should trace one of his feet on the paper, and a second student maintained that he should trace *two* of his feet. The teacher remarked that these were all good ideas and asked pairs of students to create their own paper strips composed of *five* of the king's feet (students subsequently named this a *footstrip*). The teacher's intent in suggesting the footstrip was that students would iterate a collection of units or paces rather than just single paces. The excerpt below illustrates that students drew on their prior experience of pacing as they thought about how the new tool, the footstrip, might be used.

T: The king says he wants five. "I want five in a row." How could you use something like this?

Melanie: It would work because he would put the paper in front of itself.

Hilary: This would be faster. He'd have lots more feet, and if he had lots more feet, he could just take a big step with all the feet together and there were ten. It wouldn't be just 10, 20. Each one would be the same size and you could just add them all together.

In this excerpt, Hilary anticipated that using the footstrip would be more efficient; now she could take one "big step" of five, not five little steps. More important, Hilary and others spoke of measuring now being *faster,* indicating that they anticipated curtailing counting their individual paces (i.e., "it would be faster to count by five paces rather than by one"). In this way, the students came to use a new tool, the footstrip, as a more efficient measuring device.

Although introduced by the teacher, the actual creation of the tool by the students built on their imagery of pacing. However, the teacher did not introduce the idea of the footstrip to facilitate a discussion, as she had done in the instance of the first tool. Rather, students created the footstrip, with the guidance of the teacher, as a solution to the problem offered by the king in the ongoing narrative. Eventually, students discussed the idea that their footstrips were different sizes, since some students had smaller feet than others. Thus, when the need for a standard footstrip was suggested by the children, the teacher gave each pair of students a standard footstrip.

Measuring with the footstrip—part 1

The subsequent instructional activities involved having students measure the lengths of objects on the playground with their new footstrips. Many of the students had difficulty measuring the length of an object when parts of the last iteration of the footstrip extended beyond the endpoint of the object. For example, in a whole-class setting, the teacher asked one pair of students to measure the length of a cabinet that was situated along one of the walls in the classroom (see fig. 6.2). After three iterations, the students found that there was not enough room to place another full footstrip before they reached the wall. Instead of sliding the end of the footstrip up the wall, the two students placed the end against the wall so that it overlapped the third placement of the footstrip. Then, they counted "16, 17, 18 1/4" backward along the footstrip from the wall until they reached the endpoint of the third placement. This indicates that measuring for them meant covering the space between the two ends of the cabinet exactly (i.e., no part of the footstrip could extend beyond the cabinet).

Since some students seemed to be confused by this solution method, the

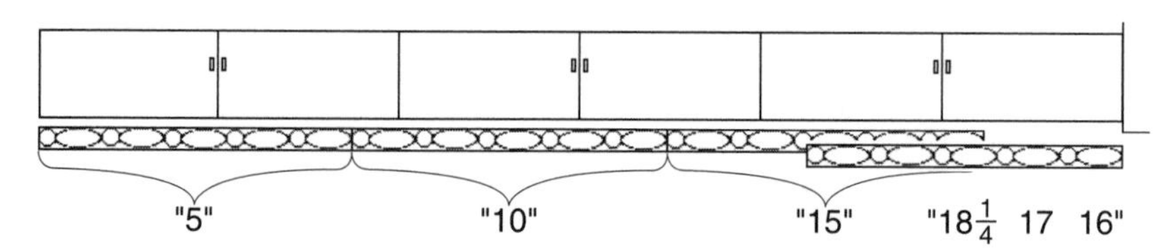

Fig. 6.2. Sandra and Porter's method of measuring the cabinet

teacher suggested that it might be easier for others to understand if one of the students doing the measuring moved the footstrip so that the excess ran up the wall. Several students rejected this suggestion. They argued that the part of the footstrip that was placed up the wall would be measuring part of the wall. The teacher asked other students to explain why they thought the footstrip could be extended up the wall. The issue was resolved when one student commented that he just pretended "in his mind" to cut the footstrip at the end of the cabinet and counted only the paces along the cabinet. The students seemed to accept this and, in subsequent whole-class discussions, imagined cutting the footstrip.

The explicit attention given to this issue can be attributed to the students' use of the footstrip. When they measured by pacing, the students could simply remove their last foot if part of it extended beyond the end of the object and refer to the extra space by pointing to it (e.g., 15 and a little bit of space). However, when they used the footstrip, *three* or *four* extra paces might extend past the endpoint rather than part of a single pace. Further, they could not resolve the difficulty by turning the footstrip sideways to "fill up" the space exactly as they had when they measured by pacing. Their use of the footstrip, therefore, made explicit the issue of what to do when a unit (i.e., a pace) did not fill up the length of an object exactly. This, in turn, led to the idea of mentally cutting the footstrip along any point as the need to do so arose.

As these examples illustrate, the teacher was successful in guiding the students' creation of a footstrip as a tool that built on their prior activity of pacing. The intent was that students would measure by iterating a collection of units, not single units. Instead of simply handing the students a ready-made footstrip, the teacher continued the narrative of the king and prompted a search for a more efficient measuring device. Through the process of creating and using footstrips, the students constructed more-sophisticated understandings of measuring. As we saw above, the fact that the footstrip was an inseparable collection of five paces led the students to develop the idea that space was something that could be partitioned by a unit or portions of a unit rather than something to be covered exactly.

Measuring with the footstrip—part 2

A second issue emerged the same day that the students measured items with their footstrips on the playground. After the students had measured an item, say 25 feet, most were not able to mentally figure out how many *footstrips* long the item was. In other words, these students found it difficult to coordinate units of different sizes. It had been our explicit goal at the outset of the teaching experiment to enable students to eventually structure quantities in various ways (e.g., 5 units of 5 paces as well as 25 paces). Hence, during a whole-class discussion the teacher asked the class how many footstrips long a cabinet was (25 feet). In order to support structuring space into different-sized units, the teacher placed masking tape at the beginning and end of each iteration of a footstrip (five altogether). The students then argued that 25 feet was the same as 5 footstrips. They reasoned that whether they measured it in feet or footstrips, the length was the same. The mathematically significant issue that arose in this conversation was the invariance of quantities of length when the unit of measure changes. This was made possible by the footstrip and the masking tape record, for students were now able to reason about the space covered by iterations of five paces rather than about single paces.

Example 3: Iterating a Bar of Ten Cubes

Eleven days after the beginning of the measurement sequence, the teacher in the classroom began a second narrative about a community of Smurfs (blue cartoon creatures) that lived in mushroom houses. In this narrative, the Smurfs encountered various situations in which they needed to measure objects and did so by placing empty food cans end to end. The students solved the problems the Smurfs encountered by using Unifix cubes as substitutes for food cans. The reason the teacher introduced a new scenario was that we hoped the students' activity of measuring with the Unifix cubes would serve as a basis for a more sophisticated measurement tool to be created later. This new scenario built on the students' prior use of their feet and the footstrip. In the king's foot scenario, the students measured with a physical extension of their bodies. In this new scenario, the students placed Unifix cubes, an "external" unit, end to end to measure lengths of objects (see also Stephan [1998] and Cobb et al. [in press]).

In the scenario the teacher developed, the Smurfs sometimes measured the lengths or heights of particular objects by stacking their food cans. Initial instructional activities included measuring the lengths of paper strips, some of which signified the length of various animals in the Smurf village. Other paper strips signified the length of the animals' pens. The students were asked to measure the pen and the length of the animals in order to find which animals fit in the pen. If an animal did not fit, students were asked to find how much longer than the pen the animal was. Typically, students snapped several cubes together to make a rod that they could extend along the full length of the object. They then counted each cube to find the length of the object.

Creation of a new tool

As the narrative of the Smurfs continued, the teacher explained that the Smurfs did not want to take an unlimited number of cans (cubes) with them when they conducted their measurements. The students suggested several alternatives, such as carrying rods of seven cans and placing the rods one after another. Other students offered varying numbers of cans that the Smurfs could carry, such as two, four, ten, and twenty. The teacher told students that these suggestions were all reasonable but that the Smurfs had decided to carry ten food cans with them. (The teacher purposely chose ten, since counting by tens supports the development of strategies for addition and subtraction.) The students named this bar of ten Unifix cubes a *Smurf bar*.

This example illustrates the teacher's effort to build on the students' contributions as they tried to solve the dilemma of the Smurfs (that of reducing the number of cans they carried for measuring). Clearly, the teacher wanted the students eventually to measure with collections of ten cubes rather than with single cubes. However, instead of telling them to do so directly, she developed a problem that required the creation of a new measurement tool. The teacher knew that a number of the students could count by tens; therefore, she was confident that she could allow a variety of students' suggestions and still have at least one student offer ten as a choice. Measuring with Unifix cubes and the Smurf bar built on the students' prior activity of measuring by pacing and iterating a footstrip. The difference was that students were counting single cubes (units outside of their bodies) and collections of ten (rather than five).

Measuring with the Smurf bar

The initial instructional activities with the Smurf bar involved measuring various items around the Smurf village (classroom). In whole-class discussions, the teacher typically asked questions that focused on the results of measuring with the Smurf bar. For example, if a student was finding the length of a table by iterating a Smurf bar and counting "10, 20, 30, 33," the teacher might ask the student "Where is 33?" or "Can you show how long something 33 cans is?" She posed these types of questions for two reasons.

First, she was trying to support the students' interpretation of measuring as an "accumulation of distance." In other words, when the students had completed a third iteration and uttered "30," some students may have reasoned that "30" signified the space covered by the last iteration (10 cubes) rather than the space spanned by all 30 cubes. It was therefore important that this issue become an explicit topic of whole-class discussion.

Second, she was trying to encourage explanations that focused on the *meaning* of measuring with the Smurf bar or food cans instead of simply describing a *method* of measuring. Questions such as "Where's the 33?" served to focus discussions not only on *how* a student obtained 33 as the measure but also on his or her interpretation of the result of measuring. As a consequence of students participating in these types of discussions, "the whole 33," for instance, came to mean the space extending from the beginning of the first cube to the end of the thirty-third cube. (See McClain et al. [1999] for a more detailed account.) In our experience, an accumulation of distance interpretation is a relatively sophisticated achievement for first graders.

As mentioned earlier, the students' use of the Smurf bar as a measurement tool emerged naturally from their measuring with the footstrip. Thus, the learning that had occurred previously as students used the footstrip was self-evident as they measured with the Smurf bar. There was, for example, no discussion concerning "filling up" the length of an object *exactly* with the Smurf bar. This issue had been discussed earlier, and students were able to build on what they had learned in these conversations to talk about more-sophisticated mathematical issues, such as measuring signifying an accumulation of distance. Again, we can see that sequencing the tools in such a way as to build on the students' experience with previously used tools supports the emergence of increasingly sophisticated mathematical conversations and, thus, learning.

Example 4: Measuring with a Strip of 100 Cans

As the teacher continued the narrative of the Smurfs, the students measured various objects that the Smurfs used or needed in their village. One of these activities involved asking students to measure and cut adding machine tape into different lengths that signified pieces of wood to be used for rafts. One pair of students, Nancy and Meagan, kept track of each iteration of the Smurf bar by writing the numerals (10, 20, 30, ...) on the adding machine tape as the two students found the lengths they needed. Building on this symbolizing activity, the teacher asked the class if they could use Nancy's and Meagan's records to think of a way to measure without having to take *any* food cans with them. Several students suggested that just as these students had marked the adding machine tape, the Smurfs could measure a piece of paper that was 10 cans long, using it instead of the Smurf bar. By posing the questions this way, the teacher elicited ideas from the students so that they contributed to the development of the new tool, a paper record of ten cans. After further discussion, the students constructed strips of paper 10 cans long, with each individual can marked (see fig. 6.3).

In a subsequent whole-class discussion, two students came to the whiteboard at the front of the classroom and showed how they would measure it with their 10-strip. They counted the strip, saying "10, 20, ..." each time, and recorded the endpoint of each iteration with a marking pen. As the students recorded each iteration, the teacher taped a 10-strip below their record of measuring on the whiteboard. After she taped each 10-strip, she asked, for instance, "Where is the 40?" This type of question further supported students' interpreting their result, 40, as an accumulation of all the distance up to 40, not just the 40th space. In this way, the teacher was attempting to guide the development of a measurement strip 100 cubes long that would be introduced in the following class period (see fig. 6.4).

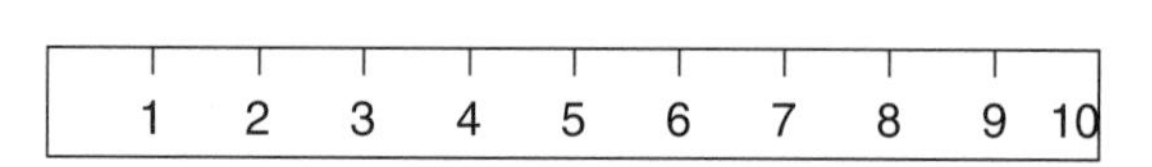

Fig. 6.3. Students' construction of a 10-strip

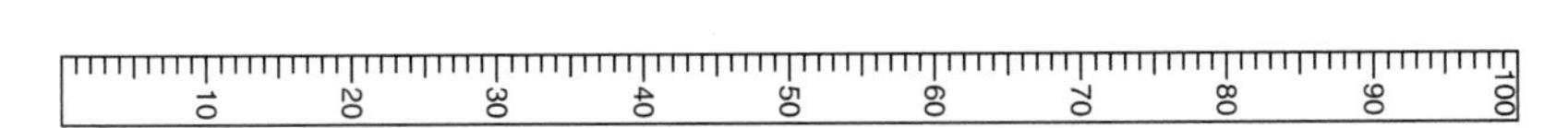

Fig. 6.4. The standard measurement strip

It could be argued that the 100-strip was imposed on the students by the more knowledgeable teacher. However, we argue that the teacher guided the development of the measurement strip by taping the 10-strips end to end as the students measured with one strip. She was attempting to ensure that the measurement strip built on students' imagery of counting Smurf bars and10-strips. The teacher also focused the discourse on the meaning the numerals had for students (e.g., "What does 40 mean?"). In this way, she was attempting to support an understanding of the 100-strip as both the result of iterating 10-strips and the result of accumulating distances.

The initial activities with the new measurement strip again involved measuring various items around the classroom. These activities seemed routine for most of the students in that they simply placed the measurement strip next to the item being measured and read off the measure of its length. Most students understood that a numeral on the strip indicated the distance from the beginning of the strip up to the numeral (an accumulation of distance). This sophisticated interpretation was made possible by the students' participation in previous whole-class discussions where an accumulation-of-distance interpretation had been an explicit focus of the conversation.

In summary, it should be clear that the measurement strip and the other tools the students used did not have any magical educational properties. Rather, each tool used in the measurement sequence was introduced in such a way as to build on the students' prior measuring experiences. For example, when the teacher taped 10-strips together, she supported the construction of the measurement strip using the students' prior experience of measuring with the Smurf bar and 10-strips. The tools used in the classroom teaching experiment were sequenced in such a way as to build on the measuring conceptions students had already constructed. Rather than beginning with the conventional ruler, the teacher built on the students' experience of pacing and counting single units and collections of units to support the eventual construction of a ruler type of tool. As a consequence, the students' reasoning when they used the ruler type of tool (i.e., the measurement strip) was extremely sophisticated.

Conclusion

In this article we have emphasized two main points. First, the introduction of tools should be sequenced to ensure that they continually build on students' prior experience with other tools. The measurement sequence was designed so that students built on their prior knowledge to create new tools and understandings for new purposes. In this way, the students always had prior experience to draw on when they used a new tool. Each tool was both a product of, and a means of, supporting further learning.

The second point we have stressed is that the teacher was able to introduce new tools that fit with her instructional agenda without stifling the students' creativity. In most examples, we described how the teacher capitalized on the students' contributions as she introduced each new device. Further, we noted that each tool was introduced as a solution to a problem. Frequently, teachers present solution methods and tools *before* the students realize there is a problem in the first place. If students do not know what the tool or a method is a solution for, they have difficulty making sense of it. In our teacher's case, she developed each problem in the context of a story scenario (e.g., the king does not want to measure everything himself, the Smurfs do not want to carry a large volume of cans, etc.), and each tool was created and used as the students addressed these problems. In our view, this aspect of "problematizing" is crucial to the development of tools as effective means of supporting students' development of increasingly sophisticated mathematical reasoning.

REFERENCES

Cobb, Paul, Michelle Stephan, Kay McClain, and Koeno Gravemeijer. "Participating in Classroom Mathematical Practices." *Journal of the Learning Sciences,* in press.

Lubinski, Cheryl. and Diane Thiessen. "Exploring Measurement through Literature." *Teaching Children Mathematics* 2 (January 1996): 260–63.

McClain, Kay, Paul Cobb, Koeno Gravemeijer, and Beth Estes. "Developing Mathematical Reasoning within the Context of Measurement." In *Developing Mathematical Reasoning in Grades K–12,* 1999 Yearbook of the National Coucil of Teachers of Mathematics (NCTM), edited by Lee V. Stiff, pp. 93–105. Reston, Va.: NCTM, 1999.

National Council of Teachers of Mathematics (NCTM). *Professional Standards for Teaching Mathematics.* Reston, Va.: NCTM, 1991.

Stephan, Michelle. "Supporting the Development of One First-Grade Classroom's Conceptions of Measurement: Analyzing Students' Learning in Social Context." Doctural diss., Vanderbilt University, 1998.

Thompson, Alba, and Patrick Thompson. "Talking about Rates Conceptually, Part II: Mathematical Knowledge for Teaching." *Journal for Research in Mathematics Education* 27 (January 1996): 2–24.

7

Promoting the Use of Diagrams as Tools for Thinking

Carmel M. Diezmann

Lyn D. English

A DIAGRAM is a visual representation that displays information in a spatial layout. In problem solving, a diagram can serve to "unpack" the structure of a problem and lay the foundation for its solution. Hence, students are often recommended to use diagrams in solving mathematical problems. For some students, this recommendation is helpful, and they are able to use the diagram as a tool in mathematical thinking and learning. However, for other students, this suggestion is singularly unhelpful. Some students are unable or perhaps unwilling to use diagrams on a problem-solving task. Students' lack of knowledge about the utility of a diagram suggests that they need help to use the diagram as an effective mathematical tool. This paper explores the knowledge that students require to become diagram literate and identifies some of the difficulties that students experience in their use of diagrams. As we discuss here, it is essential that students know *why* a diagram can be useful in problem solving, *which* diagram is appropriate for a given situation, and *how* to use a diagram to solve a problem. The paper concludes with suggestions for developing diagram literacy within the mathematics curriculum.

Diagram Literacy

The term *diagram literacy* refers to knowing about diagram use and being able to use that knowledge appropriately. The ability to use diagrams effectively is integral to mathematical thinking and learning (Nickerson 1994). Diagram literacy is a component of visual literacy, which is "the ability to understand [read] and use [write] and to think and learn in terms of images" (Hortin 1994, p. 25). Visual literacy has long been recognized as a neglected area in education (Balchin and Coleman 1965; Box and Cochenour 1994).

In the following example, ten-year-old Kate demonstrates diagram literacy (see fig. 7.1). Notice how Kate displayed the rows of the well as a vertical number line and then indicated the forward and backward movement of the frog with a series of arrows. The spatial layout of the arrows enabled her to keep track easily of the frog's movements over a number of days. Kate's diagram thus provided her with a clear representation of the problem structure and formed the basis of a successful solution.

A frog was trying to jump out of a well. Each time the frog jumped, it went up four rows of bricks, but because the bricks were slippery it slipped back one row. How many jumps will the frog need to make if the well is 12 rows high?

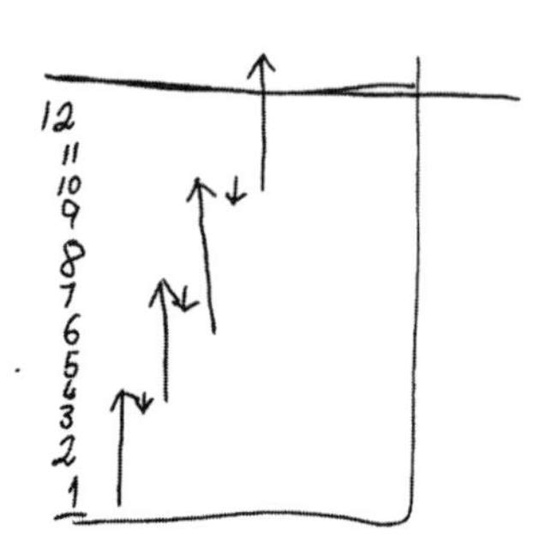

Fig. 7.1. Kate's diagram for the Frog task

In contrast, Kate's classmate, Helen, demonstrates a lack of diagram literacy on a similar task. Ten-year-old Helen's diagram is structurally inadequate because she has not incorporated a number line into her diagram and she has not represented the movement of the koala. Instead, she has focused on representing surface features, such as the branches of the tree (see fig. 7.2). As a consequence, Helen had difficulty in reasoning from her diagram and was unable to solve the problem.

A sleepy koala wants to climb to the top of a gum tree that is 10 meters high. Each day the koala climbs up 5 meters, but each night, while asleep, slides back 4 meters. At this rate how many days will it take the koala to reach the top?

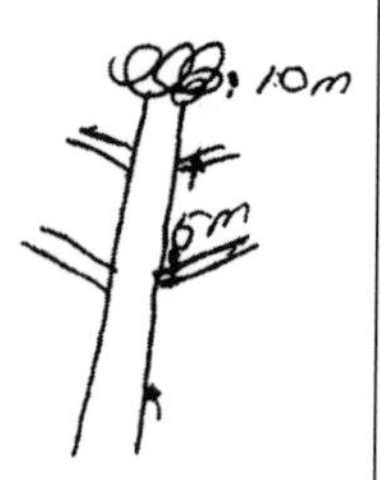

Helen: "He climbs up 5 meters to there (5-meter mark) and that took him one day and that took him back down to here (just below the 5-meter mark) and he had to climb up another 5 (meters) the next day and he got about here....

Fig. 7.2. Helen's diagram for the Koala task

Helen's difficulty in diagram use is typical of many of the students with whom we have worked. Our observations are consistent with research that has highlighted students' reluctance to use diagrams and their difficulties in doing so (e.g., Dufour-Janvier, Bednarz, and Belanger 1987; Shigematsu and Sowder 1994).

The initial step in generating an appropriate diagram is to identify which diagram is suitable for a given situation. The appropriateness of a diagram depends on how well it represents the structure of a problem. Although there can be considerable variation in the surface features of diagrams, there are basically four general-purpose diagrams that suit a range of problem situations. We now focus our discussion on those diagrams that are especially useful in elementary problem solving.

GENERAL-PURPOSE DIAGRAMS

The four general-purpose diagrams that represent specific relationships among data are *networks, matrices, hierarchies,* and *part-whole* diagrams (Novick, Hurley, and Francis 1999). We describe them briefly here.

Networks consist of sets of nodes (points) with one or more lines emanating from each node that link the nodes together, such as a map of train stations. Simple networks with few nodes and links between the nodes are sometimes referred to as *line diagrams.* Kate's diagram of the Frog task is an example of a network for a simple problem, whereas the network that Helen drew for the Birds task is a more complex representation (see fig. 7.3). Helen's diagram incorporates two sets of relationships—namely, multiples of 5 and multiples of 3—which represent the repeated visits of the two birds. Her diagram was particularly effective because of the accurate positioning of the two sets of multiples on the number line. Note that Helen was able to generate this particular network after instruction about diagram use, despite being unable to produce an appropriate diagram for the simpler Koala task.

> A robin comes to a bird feeder every 5 days and a sparrow comes by every 3 days. Today, the robin and sparrow both came to the bird feeder. How many days will it be before the robin and sparrow both come again on the same day?

Fig. 7.3. Helen's network diagram for the Birds task

Matrices use two dimensions to represent the relationships between two sets of information. Matrices are particularly useful in problems that require deductive thinking or combinatorial reasoning. A matrix is useful in deductive problems because the diagram helps the solver keep track of known information and enables implicit information to become explicit. For example, in the matrix drawn by Damien (fig. 7.4), he used the clues of "Sally and Rick met when one of them won a swimming race" and "Sally is not a swimmer" to correctly deduce that Rick was a swimmer. This fact was explicated on the matrix and represented with a check. However, Damien's reasoning was not always sound, and he also reasoned incorrectly from the clues in this problem. His difficulty is discussed shortly. In Damien's matrix, the people are represented on the horizontal axis and the sports on the vertical axis.

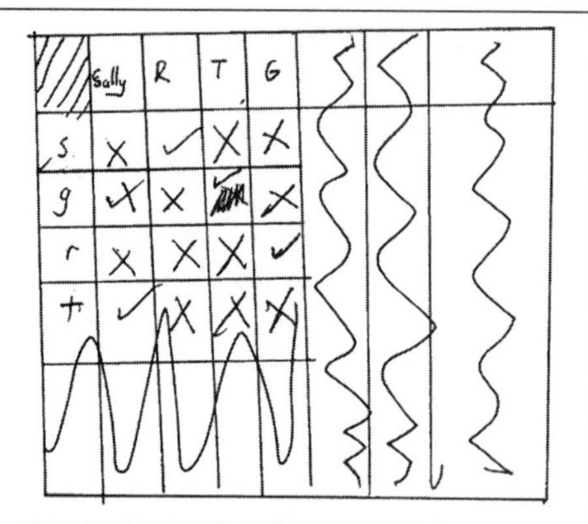

Four friends like different sports. One likes tennis, one likes swimming, one likes running, and one likes gym. Each person likes only one sport. Use the clues to help you find out which sport each friend likes.

1. Sally and Rick met when one of them won a swimming race.
2. Tara and Greg met when one of them was exercising at the gym.
3. Sally is not a swimmer or a runner.
4. Greg is a friend of the gymnast's brother.

Damien: "Aw, because it says I know that Sally's not a swimmer and a runner so I thought she was a gymnast so i ticked that and crossed them all out, and this one says that Tara and Greg met when one of them was exercising at the gym and I don't think it was her and I'm getting confused."

Fig. 7.4. Damien's reasoning for the Sports task

In a combinatorial problem, the matrix provides a visual representation of the number of combinations. For example, if children have a choice of two types of drinks and three types of food at lunch, the six lunch combinations can be seen easily on a matrix. Although some students can simply calculate the number of combinations from the number of each item, other students cannot; hence, the matrix provides an important visual referent for them.

Hierarchies comprise diverging or converging paths among a series of points. Tree diagrams and family trees are some common examples of hierarchies. Brian's simple diagram (see fig. 7.5) is an example of a converging tree diagram, which he used to represent the structure of the given problem. On the first row all the teams are represented. On the second row only the winners of the first round are shown. On the final row the overall winner of the competition appears.

Four teams competed in a knockout volleyball competition. Team A beat Team B and Team C lost to Team D. Who was the winner if Team D lost the final game?

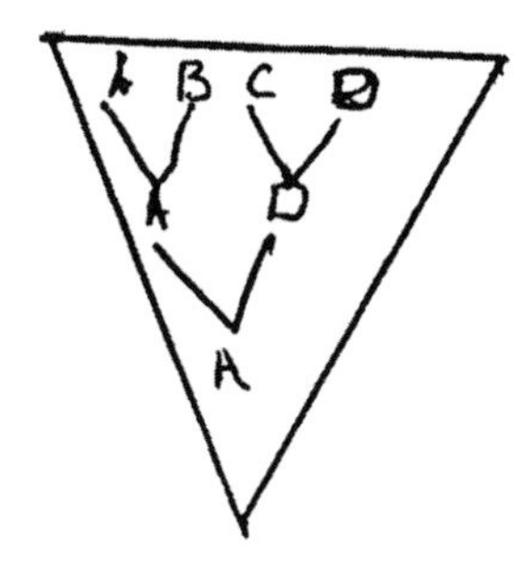

Fig. 7.5. Brian's hierarchy for the Volleyball task

Part-whole diagrams represent the relationship between a part and a whole. In contrast to matrices and hierarchies, part-whole diagrams do not have a readily recognizable external form. Damien's diagram of the Park task is an example of a part-whole diagram where the parts are the number of legs for people and dogs and the whole is the total number of legs (see fig. 7.6).

The Park: Jane saw some people walking their dogs in the park. She counted all the legs and found that there were 48 legs altogether. How many people and how many dogs were in the park? Are there any other solutions?

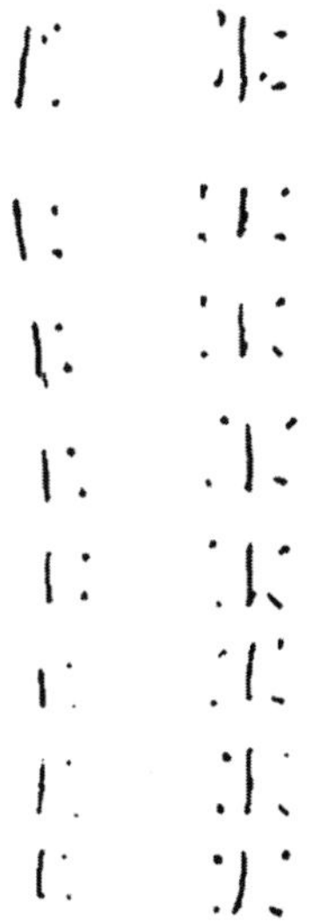

Fig. 7.6. Damien's part-whole diagram for the Park task

Students who are diagram literate are knowledgeable about each of these diagram types. They know when to use these diagrams and how to reason with them to reach a successful solution. We now consider some of the difficulties that students experience in using these general-purpose diagrams.

STUDENTS' DIFFICULTIES IN DIAGRAM USE

Students' difficulties with diagram use present obstacles to their development of diagram literacy. Diagrams do not automatically become useful tools for students. Their effectiveness may be hampered by students' lack of understanding of the concept of a diagram, their inability to generate adequate diagrams, and their failure to reason appropriately with them (Diezmann 1999). We address each of these difficulties in turn.

The Concept of a Diagram

The terms *diagram, picture,* and *drawing* are sometimes used as synonyms. However, there is an important difference between diagrams and pictures and between diagrams and drawings. Diagrams are structural representations; surface details are unimportant. For example, in Damien's diagram the people and dogs are simply represented by lines and dots (see fig. 7.6). In contrast, pictures and drawings generally show surface details. Students who are unsure of the meaning of the term *diagram* can be misled if teachers liken diagrams to pictures or drawings. This can be seen in the following interaction between a teacher and ten-year-old Frank:

Teacher: Could you draw a diagram or a picture?

Frank: What's a diagram?

Teacher: Diagrams are just like pictures. Could you draw a picture?

Frank: Um, yeh.

Teacher: Would that help?

Frank: Yeh, but it'd take ages because you'd have to draw a lot of chickens and a lot of pigs.

Frank's final response suggests that he is thinking about the surface details. We, as teachers, need to emphasize the representation of the problem structure and de-emphasize the representation of the surface features. A lack of understanding of the concept of a diagram may explain why students often represent the surface features of a problem rather than the structural features (Dufour-Janvier, Bednarz, and Belanger 1987).

Generating a Diagram

In our work, we have observed that students have varying levels of success in generating appropriate diagrams for a particular problem. Some students cannot even commence the process of diagram generation. For example, although Jon had thought of drawing a diagram, his lack of understanding of scale thwarted his attempts at doing so:

Teacher: Could you draw a picture or a diagram to show what was happening?

Jon: I'll try, but...

Teacher: You're not sure about that?

Jon: No... I thought of drawing a 10-meter tree and then each time going up 5 meters.

Teacher: Have a quick go and see if you can do that.

Jon: I don't have enough room to do a 10-meter tree (*referring to the length of his page*).

Other students can generate diagrams but are not always able to represent the problem structure adequately. Whereas Damien's part-whole diagram (fig. 7.6) indicates a sophisticated representation of the problem, Candice's and Gemma's diagrams for the same problem represent only part of the problem structure (see fig. 7.7). Candice correctly represented one dog and one person but did not represent the total number of legs, whereas Gemma correctly depicted the total number of legs but her grouping of legs was inappropriate.

Fig. 7.7. Candice's and Gemma's partially correct diagrams

Reasoning with a Diagram

Students' difficulties in reasoning with a diagram are often related to the inferences they derive about a given problem structure. We have observed difficulties occurring throughout the solution process, in particular during the generation of a diagram and when the diagram is used to reach a solution.

Inappropriate reasoning during the diagram generation process can be seen in Lisa's work (see fig. 7.8). She initially drew the three body parts separately. Lisa then measured the head, which she had drawn an arbitrary size, and adjusted the body to suit the size of the head. Although Lisa displayed an understanding of proportional reasoning in her adjustment of the body parts, her reasoning about the size of the head was inappropriate because she failed to consider the size relationship between the head and the overall length of the dog. Thus, although Lisa used the diagram to derive information, she seemed unaware that there were constraints involved in this process.

The head of a dog is half as long as its body. The tail of the dog is as long as its head and body combined. The length of the dog including its tail is 48 centimeters. How long is each part of the dog?

Teacher:	How did you know how large to make the head of the dog?
Lisa:	I just drew the head of the dog.
Teacher:	So you drew it any size and measured it?
Lisa:	Yes.
Teacher:	And it was?
Lisa:	Three centimeters.
Teacher:	How long did you make the body?
Lisa:	Six centimeters.
Teacher:	And why was that?
Lisa:	Because the head is half as long as the dog's body.

Fig. 7.8. Lisa's diagram for the Dog task

Difficulty in reasoning with a diagram to reach a solution is evident in Damien's work (fig. 7.4). Although Damien generated an appropriate matrix, his deductive reasoning was flawed because he did not consider all possibilities. Damien's explanation suggests that he had not considered that Sally might also be a tennis player.

We now offer suggestions for overcoming each of the difficulties that we have identified.

SUGGESTIONS FOR DEVELOPING DIAGRAM LITERACY

From our experiences in fostering students' diagram literacy (Diezmann 1999), we believe that specific content needs to be addressed if students are to become confident and literate in diagram use. This content should include a focus on each of the following: (1) the concept of a diagram, (2) diagram generation, and (3) reasoning with a diagram.

Understanding the Concept of a Diagram

Many students have limited or inappropriate understandings of a diagram. Teachers can help students develop these understandings in the following ways:

- Providing opportunities for students to explicate their ideas about diagrams and responding to what they say or write. Students need to develop the understanding that diagrams are representations of the structure of problems and hence differ from pictures.
- Monitoring the development of students' knowledge and understanding of diagrams over time in order to provide appropriate instruction. For example, Ian initially held the limited view that "a diagram is looking down at the location of things." However, after instruction that introduced a variety of types of diagrams, he was able to identify specific types of diagrams and when they would be useful (see fig. 7.9).

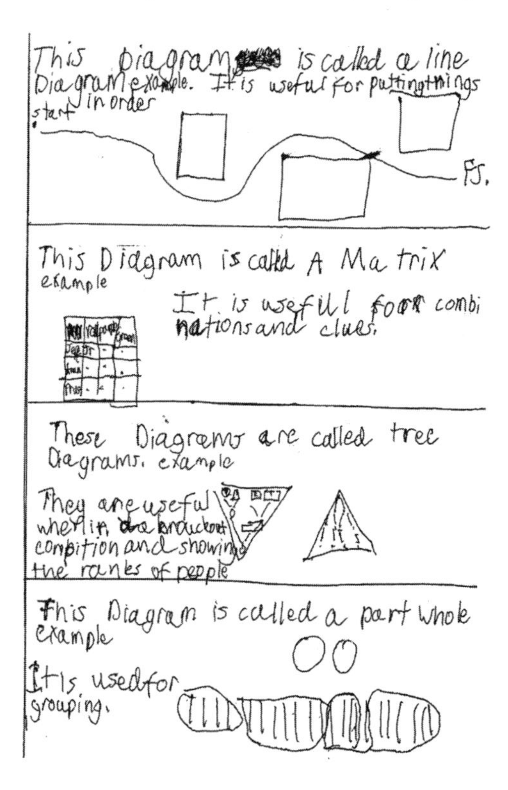

Fig. 7.9. Ian's reflection about diagrams

Generating Appropriate Diagrams

Generating a diagram is a crucial step in reasoning about a diagram (Barwise and Etchemendy 1991). During the generation process, students are afforded an opportunity to reflect on the adequacy of their diagram as a representation of the given problem. Reflection as part of the diagram generation process can enhance students' understanding of the problem structure (Nunokawa 1994). We observed such an improvement in Candice's understanding when she recognized the mismatch between the problem structure and her diagram (see fig. 7.10). Her initial interpretation would have resulted in a koala climbing into the clouds.

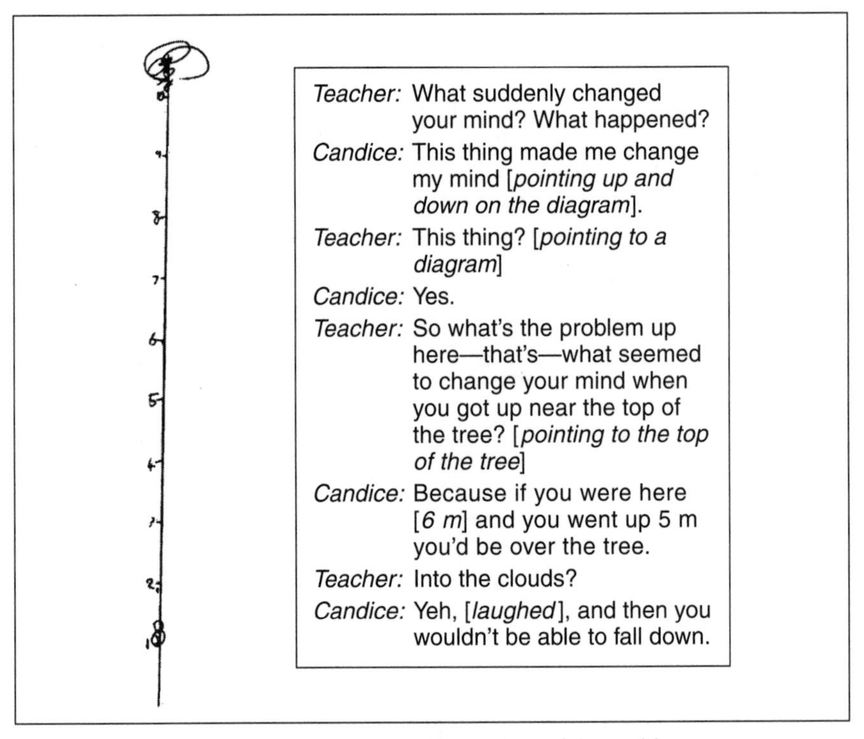

Fig. 7.10. A change in Candice's understanding of the problem structure

Teachers can support students in developing an understanding of the relationship between a problem and its representation by—

- explaining the links between the structure of a problem and its diagrammatic representation;
- modeling the generation of a diagram and explaining how the various components of the problem are represented;

- encouraging students to discuss their diagrams with one another to highlight the similarities and differences in various diagrams that may represent the same problem;
- focusing on each general-purpose diagram in turn, and presenting students with sets of problems that can be represented with a particular diagram—for example, students could be presented with a set of combinatorial and deductive problems to highlight the utility of a matrix;
- providing opportunities for students to identify which of the general-purpose diagrams is most appropriate for particular problems.

Reasoning with a Diagram

Of importance in reasoning with a diagram is the ability to make accurate inferences. However, the type of inference that can be made varies from one type of diagram to another. One of the impediments to students' reasoning is that they fail to realize that a given diagram should yield consistent answers. For example, Helen obtained three different answers from her diagram (see fig. 7.2). Initially, Helen produced an answer of three days by tracing the path taken by the koala on her drawn tree until the koala reached the top of the tree. When Helen was asked to show how she had reached the answer, she retraced her movements and gave another answer: "[It] would be four days." Helen was then asked if she could use her diagram to check her (second) answer. Helen did not respond to the question but stated a further answer: "I think it would be five [days]." During this interaction, Helen gave no indication of surprise that each time she traced the path of the koala, she reached a different answer. Helen obtained different answers because she lacked precision in her movements on the diagram.

Helen's difficulty is not uncommon, and many students become geographically lost when they traverse a diagram during the solution of a problem. Teachers can support students' reasoning with a diagram by the following means:

- Emphasizing the importance of precision in location and movement on a diagram. For example, if Helen had marked off her tree in one-meter sections, she would have been able to identify particular locations and use these measures as a number line (see fig. 7.2).
- Encouraging students to use a suitable tracking strategy and to check their work. Students can track their movement on a diagram with various indicators, such as lines, arrows, dots, or numbers. For example, in figure 7.1, Kate kept track of an animal's movements by drawing an arrow of a specific length and direction to depict each movement beside her initial diagram. Tracking indicators should be positioned to the side of a diagram rather than on top of it. Students who draw over the top of their diagrams become confused when they attempt to check their work.

CONCLUSIONS AND IMPLICATIONS

Representation is one of the ten Standards of the recent *Principles and Standards for School Mathematics* (National Council of Teachers of Mathematics [NCTM] 2000). In defining this Standard, the Council recommends that students "select, apply, and translate among mathematical representations to ssolve problems" (p. 67). We have offered some suggestions for attaining this Standard. At the same time, we have highlighted some of the difficulties that students experience in using diagrams as an effective representational tool.

Our work with students (Diezmann 1999; English 1997, 1998, 1999) suggests that we can facilitate their development of diagram literacy through—

- actively promoting the use of diagrams by modeling and discussing their use;
- emphasizing similarities and differences between problem structures;
- providing explicit instruction in the use of general-purpose diagrams and highlighting the correspondence between diagram types and problem structure;
- focusing on the diagrammatic representation of structural information in a problem rather than on its surface features;
- encouraging students to use diagram generation as a means for improving their understanding of problem structure;
- monitoring and responding to students' difficulties in the development of diagram literacy;
- ensuring that tasks are sufficiently challenging to warrant the use of a diagram.

Literacy with diagrams is an essential component of students' mathematical development (NCTM 2000). Without proactive attention from teachers, curriculum writers, and teacher educators, diagrams will remain adornments to problem texts rather than become effective tools for thinking.

REFERENCES

Balchin, William, and Alice Coleman. "Graphicacy—the Fourth 'Ace' in the Pack." *Times Educational Supplement,* 5 November 1965, p. 947.

Barwise, Jon, and John Etchemendy. "Visual Information and Valid Reasoning." In *Visualization in Teaching and Learning Mathematics,* edited by Walter Zimmermann and Steve Cunningham, pp. 9–24. Washington, D.C.: Mathematical Association of America, 1991.

Box, Cecelia A., and John Cochenour. "Visual Literacy: What Do Prospective Teachers Need to Know?" Paper presented at the Annual Conference of the Visual Literacy Association, Tempe, Arizona, October 1994. (ERIC Document Reproduction no. ED 380059)

Diezmann, Carmel M. "Assessing Diagram Quality: Making a Difference to Representation." In *Making the Difference*, edited by John Truran and Kathleen Truran, pp. 185–91. Adelaide, Australia: Mathematics Education Research Group of Australasia, 1999.

Dufour-Janvier, Bernadette, Nadine Bednarz, and Maurice Belanger. "Pedagogical Considerations concerning the Problem of Representation." In *Representation in the Teaching and Learning of Mathematics*, edited by Claude Janvier, pp.109–22. Hillsdale, N.J.: Lawrence Erlbaum Associates, 1987.

English, Lyn D. "Children's Reasoning Processes in Classifying and Solving Computational Word Problems." In *Mathematical Reasoning: Analogies, Metaphors, and Images*, edited by Lyn D. English, pp. 191–220. Mahwah, N.J.: Lawrence Erlbaum Associates, 1997.

———. "Reasoning by Analogy in Solving Comparison Problems." *Mathematical Cognition* 4 (1998): 125–46.

———. "Assessing for Structural Understanding in Children's Combinatorial Problem Solving." *Focus on Learning Problems in Mathematics* 24 (4) (1999): 63–83.

Hortin, John A. "Theoretical Foundations of Visual Learning." In *Visual Literacy*, edited by David M. Moore and Francis M. Dwyer, pp. 5–29. Englewood Cliffs, N.J.: Educational Technology Publications, 1994.

National Council of Teachers of Mathematics (NCTM). *Principles and Standards for School Mathematics.* Reston, Va.: NCTM, 2000.

Nickerson, Raymond S. "The Teaching of Thinking and Problem Solving." In *Thinking and Problem Solving*, edited by Robert J. Sternberg, pp. 409–49. San Diego, Calif.: Academic Press, 1994.

Novick, Laura R., Sean M. Hurley, and Melissa D. Francis. "Evidence for Abstract, Schematic Knowledge of Three Spatial Diagram Representations." *Memory and Cognition* 27 (1999): 288–308.

Nunokawa, Kazuhiko. "Improving Diagrams Gradually: One Approach to Using Diagrams in Problem Solving." *For the Learning of Mathematics* 14 (February 1994): 34–38.

Shigematsu, Keiichi, and Larry Sowder. "Drawings for Story Problems: Practices in Japan and the United States." *Arithmetic Teacher* 41 (May 1994): 544–47.

8

Constructing a Foundation for the Fundamental Theorem of Calculus

Marty J. Schnepp

Ricardo Nemirovsky

This paper is an analysis of a classroom discussion that occurred in the third week of a twelfth-grade Advanced Placement calculus course. It serves as an example of an activity where technology and pedagogy elicit from students intuitive, yet sophisticated, notions of the mathematics of change and variation. In this and similar activities, intuitive conceptions of mathematics of change lead to constructive discourse from which students derive computational procedures, algebraic techniques, and graphical associations that extend prior mathematical experience and challenge personal theories.

During the last ten years, mathematics teachers at Holt High School in Holt, Michigan, have been exploring a functions-based approach to curriculum (Chazan et al. 1998). One of the authors is a teacher at Holt who has explored new ways to teach calculus as part of a broader effort to investigate alternatives to traditional approaches to the teaching of high school mathematics. He joined the SimCalc project three years ago as a teacher-researcher to investigate the following research question posed by Nemirovsky, Kaput, and Roschelle (1998): How can technologies and learning environments

This research has been supported by the SimCalc project (NSF RED-9353507) and by the National Center for the Investigation of Student Learning and Achievement at the University of Wisconsin. This Center is supported under the Educational Research and Development Centers Program, PR/Award Number (R30560007), as administered by the Office of Educational Research and Improvement, or the U.S. Department of Education. All opinions and analyses expressed herein are those of the authors and do not necessarily represent the position or policies of the funding agencies. We want to thank several colleagues who provided feedback and encouragement: Dan Chazan, Al Cuoco, Tracy Noble, Beba Schternberg, Jesse Solomon, Michal Yerushalmy, and an anonymous reviewer.

change the ways students experience the mathematics of change by tapping more deeply into students' cognitive, linguistic, and kinesthetic resources? To provide a framework for how this teacher organized his course, we will elaborate on two aspects: (1) the pedagogical emergence of the fundamental theorem of calculus, and (2) the use of mathematical representations in calculus.

THE PEDAGOGICAL EMERGENCE OF THE FUNDAMENTAL THEOREM OF CALCULUS

Traditionally, calculus courses first introduce the notion of derivative, or rate of change, then integration, and finally the fundamental theorem of calculus, which establishes the connections between the first two. Usually, students are left with the formally proved statement about the inverse relationship between differentiation and integration without a clear intuitive sense of why such a relationship exists. Although students usually intuitively see adding and subtracting as "undoing" one another, most calculus students do not develop an equivalent sense of why integration "undoes" differentiation and vice versa. If integration refers to a process of accumulating a quantity, why would the rate at which it accumulates be its inverse? Or why would the area under a curve reverse the slope of the tangent to the curve? We want to build the core of calculus education, from day one to the final test, around the development of this insight. Among other implications, this means that instead of a sequence from derivatives to integrals to the fundamental theorem of calculus, we want students to learn about derivatives *in relation to* integration and vice versa:

- When students discuss integration, we want them to recognize (1) that accumulation always occurs *at a certain rate* and (2) that this rate at any given point *is* the value of the function being integrated.
- When students discuss differentiation, we want them to recognize (1) that the rate of change is *cumulative* (e.g., velocity completely determines how much more or less position is accumulated at a point) and (2) that what has been accumulated up to a certain point *is* the value of the function being differentiated.

In colloquial language, the accumulation of speed or the speed of accumulation both get us back to the quantity we started with, with the added property that in the process we lose information about a constant value, similar to how taking successive differences involves losing a constant. This paper examines the approaches that the teacher has developed to interrelate differentiation and integration in this fashion throughout his calculus course.

THE USE OF MATHEMATICAL REPRESENTATIONS IN CALCULUS

We all experience the difference between directly *recognizing* someone and *inferring* that a person must be so and so. They are not incompatible moments. As a matter of fact, it is not unusual that on "reasoning" that a person we are looking at "should" be, say, an old neighbor, suddenly we recognize the old neighbor; or, conversely, right after recognizing in a photo, for example, an aunt when she was a child, we might surmise that the photo must have been taken in a certain year or place. The same experiential distinction between recognition and inference can be made in regard to mathematical representations. As we become fluent with mathematical notations, we develop multiple and complex capacities to recognize, say, graphical shapes, forms of equations, number sequences, and so forth, and these acts of recognition become points of departure or arrival for inferences. Particularly significant in the mathematical domain is a form of recognition that Casey (1987) has called "recognizing-in." Recognizing-in is seeing something *in* something else. This is one example he reports (Casey 1987, p. 127):

> When I was working on a summer job many years ago in my hometown, my employer remarked to me one day that he recognized my father in me. When I asked him how this was so, he said that I had "my father's walk"—his very gait, his style of walking. His perceiving of my walking was imbued with remembering; or rather, his perceiving me the way he did *was* his remembering.

His employer was not recognizing him *as* his father, because he was fully aware that he was the son, not the father. But he saw his father *in* his walk, in an aspect of the son. Recognizing-in merges perception and memory: perceiving the entity in which one recognizes (e.g., the son) is also remembering what is being recognized (e.g., his father).

A major component of this teacher's calculus course was the study of motion, through which he tried to have his students come to *recognize in* graphs and functions various patterns of motion. In this domain, and probably in the use of mathematical representations in general, the type of recognizing-in that matters is one that merges perception and imagination. As opposed to Casey's example, it is not unusual that what is being recognized in a mathematical representation alludes to imagined events and entities that might be even physically impossible. This does not mean that memories are absent in mathematical recognizing-in, of course, but that these memories are to a large extent recollections of intermingled imaginary and physical worlds. We are currently examining an episode that took place in a calculus class that we observed. The students were discussing the graph of velocity versus time for an object whose graph of position versus time appears in figure 8.1.

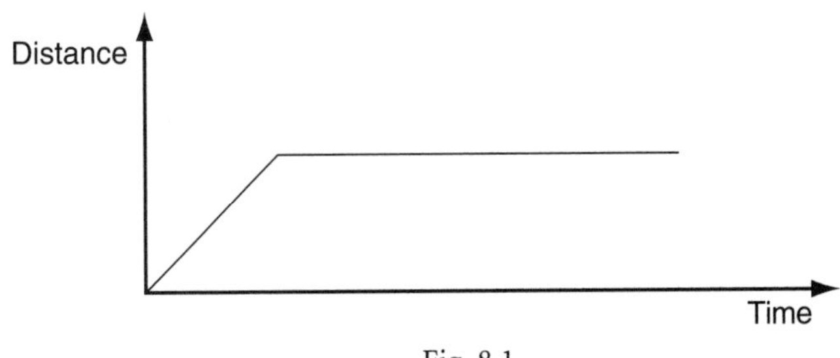

Fig. 8.1

The students discussed what to imagine and infer in order to *recognize* the acceleration of the object *in* this position-versus-time graph. The issue was the interpretation of the corner. Some of the arguments focused on whether it is physically possible for an object to stop in no time. It was obvious to them that they had never seen something stopping instantaneously, but could that happen? Other arguments questioned the relevance of this line of analysis on the basis that whether this graph represented something that was physically possible or not had no bearing on the mathematical behavior that it symbolized. There were also references to mathematical "tricks" that would make the acceleration on the corner amenable to more standard treatment. The discussion started from a direct *recognizing in* the graph of a sudden stop, which by itself was not problematic, but the question about acceleration launched the students into a complex territory in which it was essential to discriminate what was pertinent to imagine and to recall. Ultimately, what acceleration, if any, one is supposed to *recognize in* the corner was left open to further analysis.

Recognizing-in is a complex process. In Casey's example, it meant that the employer possessed a familiarity with Casey's father and an ability to discriminate perceptually ways of walking. In our mathematical examples, it involves a familiarity with physical and imaginary motion and an ability to discriminate perceptually ways of symbolizing.

Of the three characteristics of this calculus course, one of the chief ones derives from our pedagogical analysis of the fundamental theorem of calculus:

1. Differentiation and integration are introduced from a numerical perspective. They are then examined from the point of view of their mutual relationship. Rather than studying one and then the other, students encounter and then reencounter both at once (or in succession) in different contexts, levels of analysis, and representations.

The other two characteristics are based on our views about the use of mathematical representations in calculus:

2. Frequent classroom discussions focus on revising the meanings of familiar words—speed, velocity, average, distance, and so on—and

constructing mathematics related to their technical use. An ongoing aim is to realize the complexity inherent in their meanings, which are often obscured by being taken for granted (Ernest 1991). The ongoing enrichment and clarification of everyday language is a process, not of "replacing" everyday language with a technical one, but of agreeing on and refining uses and interpretations of words and ways of talking or acting

3. Tools are used that allow students to investigate physical motion and how it relates to student-constructed and formally defined functions and algorithms as well as to imaginary types of motion.

To illustrate these three traits in action, we will analyze a classroom activity and the subsequent class discussion. We will elaborate on how these features appear to facilitate a classroom discourse from which students derive computational procedures, linked graphical representations, and algebraic techniques. Other calculus educators are pursuing related approaches (Speiser and Walter 1996; Rosenthal 1992). Through this process, learners extend prior mathematical experiences, challenge personal theories, and develop sophisticated images of rate and accumulation. We will briefly describe the technology and the curricular content that generated the background for the selected classroom episode.

TOOLS: LINE BECOMES MOTION; MOTION BECOMES LINE

A computer-based pedagogical environment has been developed at TERC for the study of the mathematics of change. The tools consist of graphing software and hardware that link a computer to external physical devices. The software allows for two representational orientations, or vantage points, from which users may study the relationships between physical motion and motion graphs. These are referred to as "line becomes motion" (LBM) and "motion becomes line" (MBL). With LBM, a user constructs a graph on a computer, which in turn communicates with a motor that moves the mechanical device according to the graphical specifications.

The devices discussed in this paper are a pair of miniature cars on parallel linear tracks and a miniature stationary bike rider (see figs. 8.2 and 8.3). For these LBM devices, a user can draw or symbolically define a position-time, velocity-time, or acceleration-time graph on the computer. The software will then move the external device according to the characteristics of the user's graph. If the device is minicars, one or more cars will travel down the track moving forward or backward, speeding up or slowing down, and so on, as dictated by the graph. If the device is the minibiker, the rider's total number of pedal rotations, pedaling rate, or change in pedaling rate can be dictated by graphs.

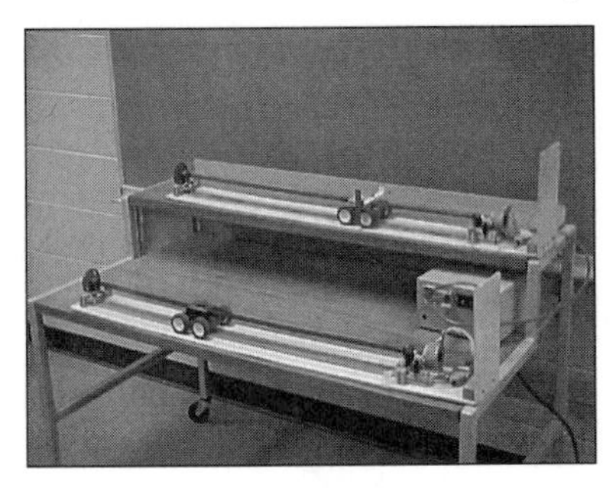

Fig. 8.2

With MBL and common modeling tasks, the motion is the object of study and known mathematical tools are used to construct a graphical or symbolic representation of it. LBM shifts the usual representational paradigm. The user-created graphs drive the motion, and thus the cars' or biker's motion becomes a representation of the graphs, and the mathematical function becomes the object of study. The impact of this shift on classroom conversations is subtle but significant.

Two aspects of the learning process are distinctive in LBM. First, LBM tends to reveal that the formal syntax of symbolic expressions has causal effects (e.g., position and velocity graphs correspond to each other if they generate the same motion). Second, device motion provides a tangible means of testing students' conjectures, interpretations, and calculations. Threatening aspects of classroom discourse diminish for many participants when an inanimate machine is validating or refuting students' ideas. We find students more willing to share and discuss ideas and the flawed or valid reasoning behind them when the machine is passing judgment rather than the teacher. When an idea is incompatible with observed motion, the student must reexamine his or her own thinking to resolve the incongruity. Classroom activities using LBM lend themselves readily to exploratory discussions in which students assume greater independence of the teacher than they ordinarily do. We have found that LBM allows us to encourage independent and group thinking and enables us to take on the role of discussion facilitators rather than conjecture evaluators. The use of LBM devices enables individual students to take increased advantage of their experiential resources involving motion and thereby to gain a deeper understanding of calculus. We have documented how small-group and whole-class conversations around core ideas in calculus can be made more productive by the use of LBM devices.

Fig. 8.3

SELECTED CLASSROOM EPISODE

The calculus program from which the following classroom episode comes has evolved over seven years from a traditional format. No textbook is used for the first semester. The opening topic of the course is numerical integration. Students discuss situations for which they are given numerical data or symbolic function rules describing the rate at which some quantity changes. They are asked to determine how much of the changing quantity has accumulated over various time intervals. Contexts using LBM and other situations ground all conversations in the early stages of the course.

Given the continuing tension in the U.S. mathematics community between skills and conceptual understanding, it is relevant that although the course's primary focus is on understanding the fundamental relationships between rate and accumulation in contexts and exploring student constructions rather than on covering the traditional list of symbolic techniques and topics, the instruction has maintained a goal of preparing students for the College Board's Advanced Placement (AP) exam. Therefore, most (but not all) of the structures and symbolic techniques of a traditional calculus class receive attention during the school year; in addition, students do perform well on the AP test. Over the seven years, 158 students have taken the test, with 125 (42 percent female and 58 percent male) earning scores of 3 or higher, which carry recommendations for college credit. Of the 36 students enrolled in the year when data were collected, 28 chose to take the AP test, with 24 earning a score of 3 or higher (15 students earned a 3, 4 earned a 4, and 5 earned a 5).

This classroom episode occurred in the first unit of the course. Students had been in class for three weeks and were developing methods for numerical approximation. They were given information about the pedaling rate for the LBM mechanical biker and were asked to estimate the total number of pedal rotations made during a time interval. They found a natural transition from previous work with constant rate problems in an approach by which they chose one rate to represent intervals in which the rate actually varied. Two of the methods students invented for choosing a rate for a given interval involved averaging rate function values. One called for adding a selected number of rate function outputs, dividing the sum by the number of values, and then multiplying this "average rate" by the time duration in order to estimate the total number of rotations. The other method was analogous to a "trapezoidal approximation." Students subdivided the time interval into subintervals. For each subinterval, they averaged the rate function's output at the beginning and the end, then multiplied this "average rate" by the subinterval's time in order to estimate the number of rotations that occurred during the subinterval. Students then added all the subinterval estimates to arrive at an overall estimate for the total number of rotations.

One of the authors was a coteacher of the class. He observed that students were using the term *average rate* in ways that held little significance beyond the "average" process. In addition, they needed to make a distinction between their use of "average rate" and the more common textbook use of that term, which is reserved for the calculation

$$\frac{f(b) - f(a)}{b - a},$$

given a function $f(x)$ on an interval $a \leq x \leq b$. He posed a problem designed with the purpose of engaging students in conversations that would establish underpinnings for the average value of a function, typically defined as

$$\frac{1}{b-a}\int_a^b f(x)dx.$$

Thus, during the third week, the teacher gave students the problem shown in figure 8.4.

Minicar Average Velocity

Consider the velocity function for the red minicar: $v(t) = 0.1(t - 11)^2 - 2$, $0 \leq t \leq 20$, where $v(t)$ is the minicar's velocity in cm/sec and t is the time in seconds.

1. Determine where the minicar is by the end of the 20-second time interval.
2. Determine a value to represent the "average velocity" of the minicar.
3. Explain what this number means in relation to the minicar's trip.
4. Make an accurate table and graph showing ordered pairs $(x, p(x))$ where $p(x)$ is an approximation of the minicar's position at any time $t = x$. Be sure to indicate the method you used and the subinterval size, Δt, that you chose.

Fig. 8.4

Students were given the Minicar Average Velocity exercise as homework at the end of a class period, two days before the session discussed here. The day following the initial assignment, students used the LBM software and watched a minicar moving according to the velocity function before working the problem further in small groups. Near the end of that period, the class reconvened to discuss answers to #2 in the problem. Solutions ranged from 1.4 to 2.7. (It may help the reader to know that the average velocity is $1.4\overline{33}$ and the average speed is approximately 2.63.)

The ensuing discussion became mired in efforts of individual students to convince others of the correct way to solve #2. The debate consumed the remaining class time, without reaching a resolution. From the students' use of the terms *speed* and *velocity*, the teachers theorized that students were quibbling over how to calculate "average velocity" when they had yet to establish what the phrase meant. The terms *speed* and *velocity* had been introduced two weeks earlier in a physics course that many students were taking concurrently. However, their interpretations of these terms were not carefully thought out and were often inconsistent. One of the teachers, recalling an activity he had observed while visiting TERC, suggested a lesson that he hoped would invigorate the conversation and elicit an awareness of the dissonant conceptions of *speed* and *velocity*.

The teachers began the next lesson with two minicars at the front of the room, positioned on parallel tracks on a two-tiered table so that students were able to see both cars. Without referring to the previous activity or discussion, the teachers posed the question: "If the red car travels according to $v(t) = 0.1(t - 11)^2 - 2$ and the blue car moves at a constant rate, how fast will the blue one have to go in order to start and stop in the same place?" Students worked collaboratively for about ten minutes and then were called together. At first, the extent to which students associated this activity and the previous problem was unclear. But as the conversation progressed, it became evident that many students were drawing on previous calculations from the Minicar Average Velocity problem.

From several suggestions, the class chose a value of 1.5 cm/sec. The red car was set to run at a velocity determined by $v(t)$ and the blue car was set with a constant velocity of 1.5 cm/sec. The graphs shown in figure 8.5 were projected on a television monitor. A time cursor moved over each graph as the minicars made their trip. At first, the test was inconclusive because the cars' initial positions were not the same. A student went to the front of the room and made several attempts to keep a ruler aligned with the rear axles of the cars as they moved. The student reasoned that if the cars "moved the same," the angle of the ruler would be the same at the beginning and end of the run. The truth of this proposition proved too difficult to determine, so the discussion continued. Students revisited the terms *displacement, posi-*

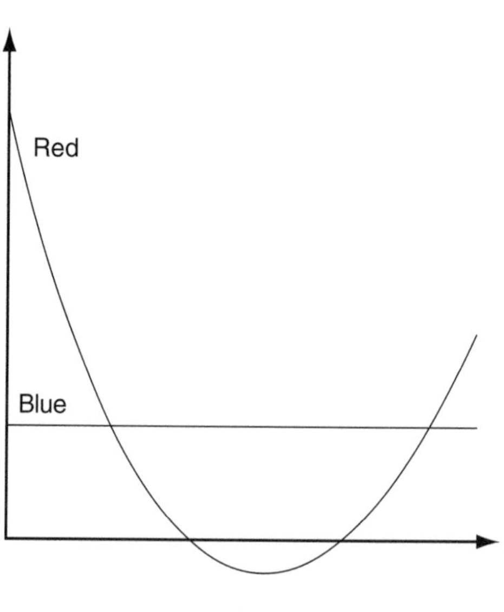

Fig. 8.5

tion, and *distance* [traveled] while a student wrote definitions on the board: "*Position*—where it is"; "*Displacement*—how far it is relative to where it started"; "*Distance*—how far it moved all together (forward and backward)." Eventually, another student suggested resetting the cars' initial positions to zero with the LBM software. This was done, and the students saw that the 1.5cm/sec rate for the blue car made the cars stop in approximately the same place.

One of the teachers asked, "So is this the idea that you guys were talking about yesterday when you were talking about average velocity?" After several affirmative responses, he asked, "So what's a good definition?" Another discussion of the meanings of *speed* and *velocity* began. A student suggested a definition of *average velocity,* reading from a paper she had apparently written earlier. Another student wrote the definition on the board. After several word substitutions suggested by other students (*rate* for *speed, object* for *minicar,* and *displacement* for *distance*), the final form of the student's definition of average velocity was as follows: "The rate that the object would have had to had constant to achieve the same displacement in the same amount of time."

This formulation prompted the following exchange between two students who began to question the sign of rate values:

> *Sam:* I'm having trouble with this negative velocity stuff. See, for this definition ...if the car were to go like 5 cm/sec forward, and 5 cm/sec back for the same amount of time and got back to where it started, we would be saying that its average velocity was zero....

The student went on to say that negative velocity signifies negative direction, and he concluded that direction was not relevant to the average. Thus, he thought that for his example the average velocity should be 5 cm/sec.

> *Sue:* So are you basically trying to say that you'd want the average velocity to equal out to the same distance in the end...same distance traveled? ...Because then you'd ignore the direction of the velocity and make everything positive.... I did that. (Both students, along with others in the class, had calculated values between 2.4 and 2.6 for #2 on the Minicar Average Velocity problem.)
>
> *Sam:* That would make more sense to me.... I think what we have here [in the definition] is average speed...for the displacement.

One of the teachers asked another student, who had worked with Sam, to come forward and draw on the board the graph that he had drawn on his paper after "making everything positive." The student drew the graph shown in

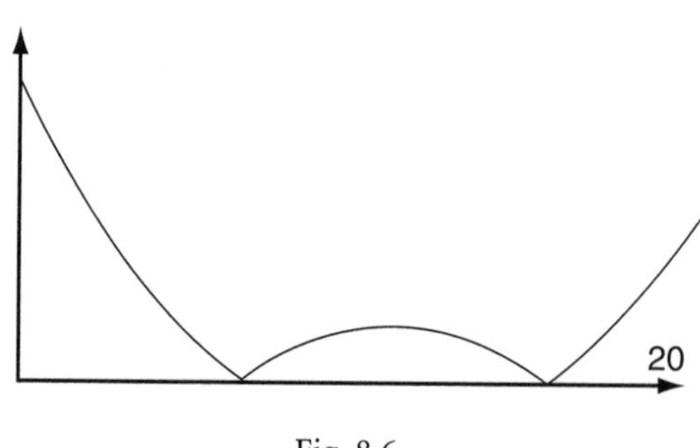

Fig. 8.6

figure 8.6, saying, "I just did absolute value.... It would cover the same distance.... It wouldn't end at the same spot in the same time. It would cover the same amount of distance."

At this point, all the students agreed that a rate between 2.4 and 2.6, if maintained for 20 seconds, would cause the blue car to travel the same total distance as the red car but not to end in the same place. Students further agreed on the significance of two distinct concepts of *constant rate*: (1) a constant rate that would give the same displacement (or change in position) and (2) a constant rate that would give the same total distance. However, more discussion of the distinction between speed and velocity followed. Students considered which quantity cannot be negative. Eventually, one student suggested a distinction between *speed* as "distance over time" and *velocity* as "displacement over time." Another student argued the opposite. The first student countered that the $|v(t)|$ graph (shown in fig. 8.6) would result in an increasing change in distance for the "reflected region," whereas the $v(t)$ graph (shown in fig. 8.5) would "accumulate a decrease" in the region where $v(t)$ had negative values. This was because the minicar moved backward during that time period. A third student pointed out that a car's speedometer cannot have negative values, so she believed that speed is always positive. This quelled the debate, though it did not convince everyone. One of the teachers redirected the class to the student's comment about accumulation.

He asked students to sketch "accumulation graphs" from the graphs of $v(t)$ on the computer and $|v(t)|$ on the board. Taking cues from earlier work, the teacher drew rectangles under the graph on the board, as shown in figure 8.7, to represent intervals assumed to be of constant velocity. A student went to the board and, with help from classmates, roughly sketched the graph shown in figure 8.8. Other students contributed comments to clarify the

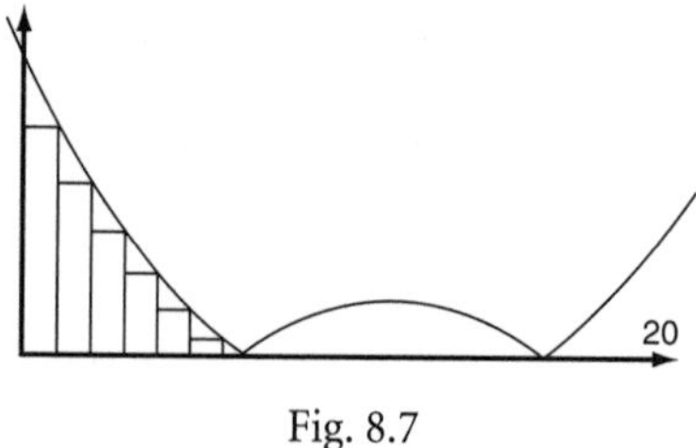

Fig. 8.7

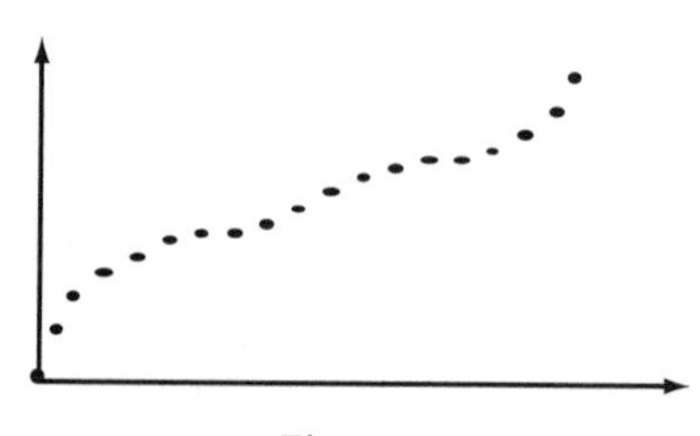

Fig. 8.8

effect of small increases near the intercepts of $|v(t)|$. One volunteered that "if you are going slow, you're not going to go as far. Since you started going slower, you're not going as far...so you're not going to have as much of an increase in total distance."

A conversation involving additional students then turned to the relationship between the areas of the rectangles shown in figure 8.7 and the changes represented by the dots in the accumulation graph shown in figure 8.8. One student explained how she thought about increase in the accumulation graph: "I said that, um, isn't it like the area of the rectangle...area under the column actually,..." One of the teachers colored one rectangle.... in the graph shown in figure 8.7 and asked how that particular object related to the accumulation graph. The student replied, "You add, um, however much area was added on...the difference between the third and fourth point...the y amount between the third to the fourth point." The teacher then labeled Δy on the accumulation graph represented in figure 8.8.

Students left class with an assignment to sketch an accumulation graph for $v(t)$. The following day, all four graphs, showing speed, total distance, velocity, and displacement, were discussed. All these subtle distinctions became part of the consensus eventually reached on the canonical distinctions between average speed and average velocity.

CONCLUSION

In elucidating how technologies and learning environments can alter the ways in which students experience the mathematics of change, we focus on the three aspects highlighted in the introduction. By using these aids to tap into students' cognitive, linguistic, and kinesthetic resources, teachers can achieve the following goals:

1. Introduce differentiation and accumulation simultaneously. The student's comment that the negative velocity "accumulates a decrease" suggests the analytic capacity to conceive accumulation and differentiation as two inextricably linked operations; "accumulate a decrease" refers to both. The description from another student of how "you're not going to have as much of an increase in total distance" (for a small velocity) also merges both aspects in a single description. Within the first month of a calculus course, students are able to develop and analyze numerical methods for approximating accumulations (integrals), relate these to the area under a graph, and begin to discuss the calculation of instantaneous rate. The relationships between the rate and accumulation calculation processes become explicit, setting the stage for a yearlong study of calculus.

2. Revise the meanings of familiar words.For example, a student's comment on his "trouble with this negative velocity" led the class to articulate a

consistent distinction between average velocity and speed, and another's comment about the "reflected region" helped them to distinguish $v(t)$ and $|v(t)|$. Through this classroom conversation, the students were not introduced to technical terms from *without* but refined their fluent use of everyday language to gain precision and logical consistency.

3. Investigate physical motion as a representation of mathematical functions. Typically, a mathematical function serves as a model of some physical phenomenon. LBM reverses the usual representational paradigm, creating unusual opportunities for mathematics learning. For instance, the definition of $v(t)$ for the minicars left undefined their start positions. The question was how to compare the two cars. Eventually, the students came to the conclusion that one had to define the start position arbitrarily and independently. This was a major insight in relation to the fundamental theorem of calculus: The cars represent the velocity functions equally well, whatever their initial positions are. This asymmetry between rate and accumulation became evident from the fact that the functional definitions left open countless possibilities in the physical realm.

REFERENCES

Casey, Edward. *Remembering: A Phenomenological Study.* Bloomington and Indianapolis, Ind.: Indiana University Press, 1987.

Chazan, Dan, David Ben-Chaim, Jan Gormas, Marty Schnepp, Mike Lehman, Sandy Bethell, and Steve Neurither. "Shared Teaching Assignments in the Service of Mathematics Reform: Situated Professional Development." *Teaching and Teacher Education* 14, no. 7 (1998): 687–702.

Ernest, Paul. *The Philosophy of Mathematics Education.* London: Falmer Press, 1991.

Nemirovsky, Ricardo, James Kaput, and Jeremy Roschelle. "Enlarging Mathematical Activity from Modeling Phenomena to Generating Phenomena." *In Proceedings of the 22nd Annual Meeting of the International Group for the Psychology of Mathematics Education,* vol. 3, edited by Alwyn Olivier and Karen Newstead, pp. 287–94. Stellenbosch, South Africa: Program Committee of the 22nd PME Conference, 1998.

Rosenthal, Bill. "Discovering and Experiencing the Fundamental Theorem of Calculus." *Primus 2,* no. 2 (1992): 131–54.

Speiser, Bob, and Chuck Walter. "Second Catwalk: Narrative, Context, and Embodiment." *Journal of Mathematical Behavior* 15, no. 4 (1996): 351–71.

9

Mapping Diagrams
Another View of Functions

Mark Bridger

Maxine Bridger

THE widespread use of the graphing calculator to "plot functions" can have the unfortunate side effect of obscuring the function concept itself. Identifying a function with its graph replaces the dynamic idea of a rule, procedure, or mapping with a static, automatically drawn picture. One way of preventing this is by introducing mapping diagrams as a simple, and supplementary, way of visualizing functions.

Some Background

Most of us who study and use mathematics have developed more than one way of thinking about a function. Leibniz originally used the term to refer to the relation between ordinate y and abscissa x that one obtained by traversing a curve (Calinger 1995, p. 387). This is still one of our main associations; we say, almost automatically, "y is a function of x" when we mean that the quantity y somehow *depends* on x. We also continue to associate this dependency with the x- and y-coordinates on a graph. Leibniz's contemporary, Johann Bernoulli, defined a *function* as an expression made from a variable by combining it algebraically with various constants (Boyer 1989). This also corresponds to a modern association: the function as formula. Thus, for example, we speak of the "function" $x^2 - 2x + 12$.

In the middle of the eighteenth century, Euler introduced the notation $f(x)$ to express a function defined using the variable x; we still commonly say, for example, "consider the function $f(x) = x^2 - 2x + 12$." At this point, the process of abstraction began, and the *name* f of the function became separated from its value $f(x)$. Most of us have seen or used some version of the following definition:

DEFINITION: *A function f is a rule or procedure that associates with each (allowable) number x a well-defined number called f(x).*

Beginning with Descartes and Fermat (Katz 1998), it became standard to represent this functional association by plotting the ordered pairs $(x, f(x))$ of the *graph* of f. This technique, however, is pedagogically useful only in those instances where the graph can actually be constructed and where the student understands what he or she is seeing. Students quite commonly fail to make the connection that the ordinate y of a point (x, y) on the graph is actually the functional value $f(x)$. In calculus courses, students often find the notation $(x + \Delta x, f(x + \Delta x))$ confusing because they have not yet constructed this connection.

When we teach a course that uses functions that either involve many variables or take as values points or vectors in 3-space, students have a great deal of difficulty grasping what the function actually *is*. For example, in "parametrizing" curves in 2- or 3-space, students are constantly looking for a graph and are very uncomfortable with seeing the curve as the *image* of a function.

In addition to these problems, dependence on the graph inhibits the development of the idea of function as *mapping*. Here the view is that a function takes a set of points—say, a line or curve or surface—and moves, associates, or *maps* it *onto* another set of points—say, another curve or surface—called the *image* of the original set of points. This idea of mapping comes from the venerable science of map projection, in which a portion of the globe is rendered on a flat piece of paper.

For example, in *stereographic projection* (fig. 9.1), the earth is viewed as a transparent sphere. A "light" is placed at the North Pole, N, and a flat screen at the South Pole, S. The rays of light project points P of the globe (with the exception of the North Pole itself) onto points Q of the chart. The image of this map is the entire plane; the equator, for example, gets mapped onto a circle.

MAPPING DIAGRAMS

To avoid some of the problems of graph dependency and emphasize the association $x \rightarrow f(x)$, we have introduced mapping diagrams as a supplementary topic. Constructing a mapping diagram is easy. Draw a vertical axis, called the *domain axis*. To the right, draw another vertical axis called the *range axis*. For some

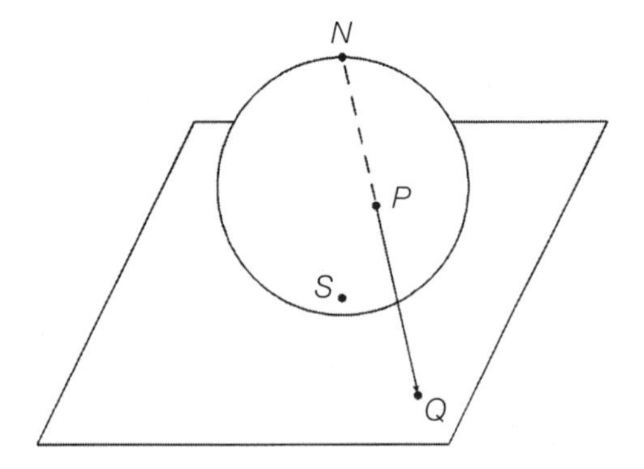

Fig. 9.1. Stereographic projection

number (point) x_1 on the domain axis, locate the point $f(x_1)$ on the range axis, and connect x_1 with $f(x_1)$ by a straight line. This is called the *mapping line* for x_1. Now construct the mapping line for another point x_2, then another for x_3, and so on. The collection of mapping lines you get is called a *mapping diagram* for the function f. (See fig. 9.2.)

Drawing mapping diagrams by hand is not a difficult chore when there are not too many mapping lines to be displayed. For more extensive diagrams, there is public domain software available to do the job—more on this later.

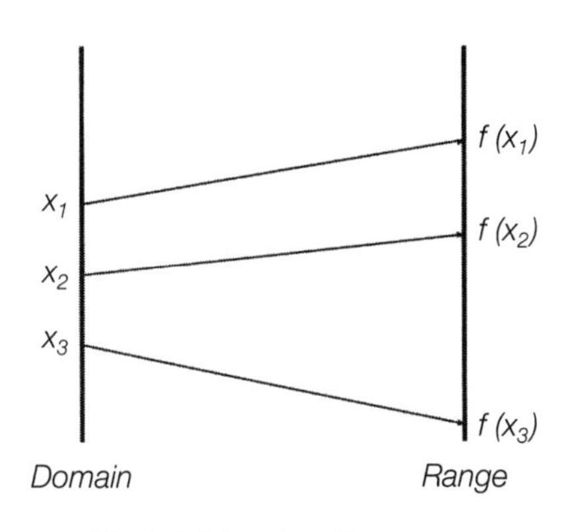

Fig. 9.2. Mapping diagram

Cartesian graphing of functions is particularly good for visualizing extrema, convexity, and asymptotes and for determining regions of increase and decrease. Mapping diagrams, as mentioned above, are useful in promoting the function-as-mapping concept. They excel in determining where a function is an expansion or contraction and where and how a mapping is a single or multiple covering of regions in its range, that is, whether it is one-one or many-one. They also are particularly well suited for visualizing compositions and inverses of functions.

In discussing these ideas, we will try to get away from the static notation $f(x)$ = (some formula) and make use of a more dynamic symbolism. For example, instead of denoting a function in the standard way—say, $f(x) = 3x^2 - 2x + 7$—we will write $f: x \rightarrow 3x^2 - 2x + 7$ or simply $x \rightarrow 3x^2 - 2x + 7$.

EXPANSIONS AND CONTRACTIONS

We begin by constructing mapping diagrams for the two functions $x \rightarrow 2x$ and $x \rightarrow 1/2x$. For these examples, it is important to use the same scales for both domain and range. In figure 9.3, we have drawn examples of such diagrams; students can easily draw them by hand. Since the limits are –4 to 4 for both domain and range, some mapping lines for $x \rightarrow 2x$ are not drawn, since they don't fit in. Note that the domain points are chosen in an equally spaced way; here they are drawn 0.5 apart.

We might ask the students, "What do you notice about the spacing of the points?" It should be eventually clear that equally spaced points in the domain lead to equally spaced image points in the range. What about actual spacing? The students soon conjecture that in the first function, the distance between image points is *double* the distance between domain points. For the second function, the distances are *halved.*

Can this be proved? If your class is ready for proofs, the following outline can be used (and, of course, generalized):

1. Distance is the absolute value of difference.
2. If u and v are domain values, their distance is $D = |u - v|$.
3. The images of u and v are $2u$ and $2v$.
4. Image distance $= |2u - 2v| = 2|u - v| = 2D$.

A similar argument works for $x \rightarrow 1/2\,x$. Stronger students can generalize to $x \rightarrow ax$ or even $x \rightarrow ax + b$: any *linear* function. The benefits of going through this demonstration are that, first of all, it is easy, and second, it shows students that proofs occur outside of geometry!

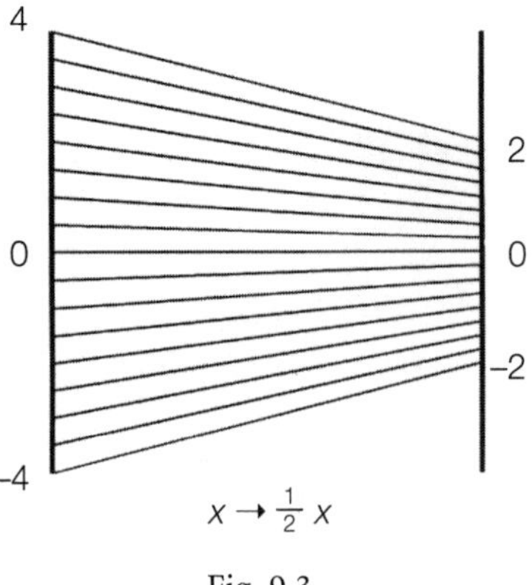

Fig. 9.3

Figure 9.4 shows what the situation looks like when the coefficient a is negative.

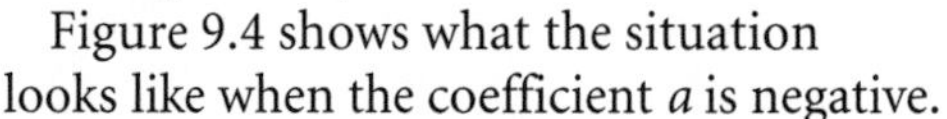

Note that although the domains and ranges are the same for these functions, the mapping lines cross each other. This is because the functions in this case are *decreasing;* points lower in the domain want to go up, points

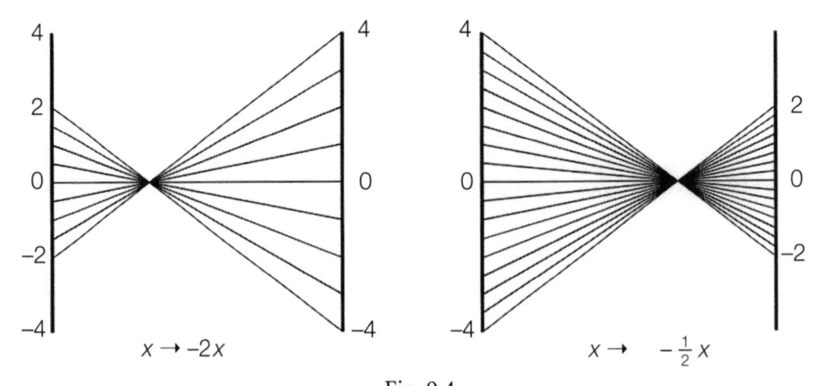

Fig. 9.4

higher in the domain want to go down, so their paths must cross. In the case of linear functions of the form $x \rightarrow ax + b$, the mapping lines must, in fact, all cross at a single point, called the *focal point.* Because the functions shown have negative slope, the focal point lies between the domain and range lines; when the slope is positive, the focal point will lie either to the right or left of these lines. The idea that focal points exist for linear functions is an interesting exercise for strong classes; a proof is included in the appendix at the end of the article. Proving the converse—namely, that a function whose mapping diagram has a focal point must be linear—is a challenging student project.

Maps such as $x \rightarrow 2x$, which move points further apart, are called *expansions.* Maps like $x \rightarrow 1/2x$ move points closer together and are called *contractions.* Here is a class question, open to discussion and experimentation: When is $x \rightarrow ax$ (or $x \rightarrow ax + b$) a contraction? (Answer: when $|a| < 1$.)

Consider the example of the Fahrenheit-Celsius conversion. We know that 0° Celsius (C) corresponds to 32° Fahrenheit (F), and 100°C corresponds to 212°F, since they represent, respectively, the freezing and boiling points of water. (See fig. 9.5 for a mapping diagram of this correspondence.)

Since 100 degrees get mapped onto 180 degrees, and we expect that equal distances get mapped into equal distances, each degree Celsius is 180/100 = 9/5 degrees Fahrenheit. Thus: $F = 9/5C$ + (constant); since $0 \rightarrow 32$, the constant must be 32, so we get the familiar formula shown in figure 9.5.

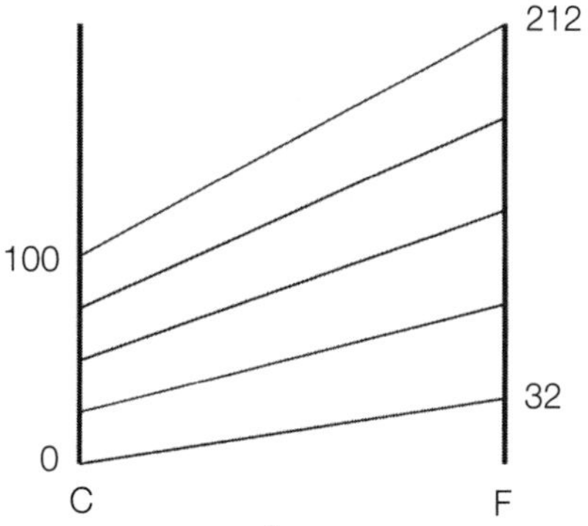

$$F = \frac{9}{5}C + 32$$

Fig. 9.5

A NONLINEAR EXAMPLE

We have been looking at linear functions, those of the form $x \rightarrow ax + b$. They have the same contraction-expansion property everywhere. Now let's look at the squaring function: $x \rightarrow x^2$. With the aid of a calculator or even hand multiplication, students can draw a simple mapping diagram for this function—say, with domain and range equal to the interval [0,4]. (Here it may be advantageous to hand out a computer-generated mapping diagram, similar to the one in fig. 9.6, for greater accuracy.)

The first thing that students notice is that equal distances do not go into equal distances; the interval [0,1] maps to the interval [0,1], but the interval [1,2] maps to [1,4]. Is this mapping at least always an expansion? Careful examination shows that it is not! In fact, the interval [0,1/4] has length 1/4, but its image is the interval [0,1/16], which is only one quarter as long.

So here is an example of a function that is an expansion in some places, a contraction in others. The natural questions are, Where is $x \to x^2$ a contraction and where is it an expansion? Many mathematics teachers and professors, when asked this question, immediately guess that $x = 1$ is the cutoff, since the function is an expansion from 1 on out. But observe: $[3/4,7/8] \to [9/16,49/64]$. The first interval has length 1/8, but the second has length $13/64 > 8/64 = 1/8$; so squaring is not a contraction on $[3/4,7/8] \subset [0,1]$.

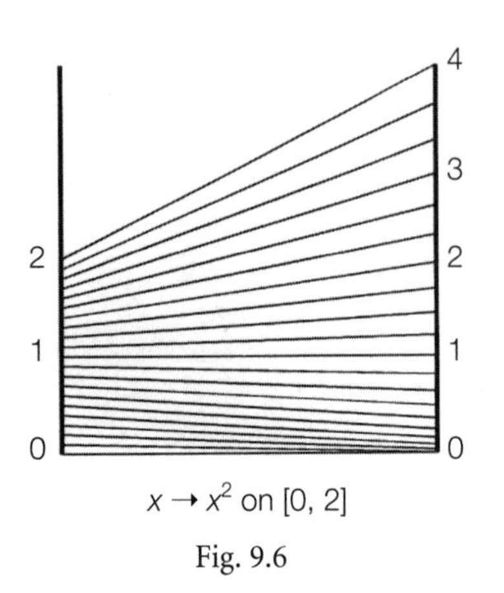

Fig. 9.6

Let's now invoke the "rule of three": analyze the problem graphically, numerically, and analytically. A calculator, spreadsheet, or computer algebra system can produce table 9.1, on which the previous mapping diagram is based.

TABLE 9.1

x	x^2	Change in x^2
0.0	0.00	—
0.1	0.01	0.01
0.2	0.04	0.03
0.3	0.09	0.05
0.4	0.16	0.07
0.5	0.25	0.09
0.6	0.36	0.11
0.7	0.49	0.13
0.8	0.64	0.15
0.9	0.81	0.17
1.0	1.00	0.19

x	x^2	Change in x^2
1.1	1.21	0.21
1.2	1.44	0.23
1.3	1.69	0.25
1.4	1.96	0.27
1.5	2.25	0.29
1.6	2.56	0.31
1.7	2.89	0.33
1.8	3.24	0.35
1.9	3.61	0.37
2.0	4.00	0.39

We see from this table, where the gap between the x's is 0.1, that the gap between their squares is less for $x < 0.5$ and greater for $x > 0.5$. This suggests that $x = 0.5$ is the cutoff between contraction and expansion for this function. This is further confirmed by the table of values in table 9.2 (where x is incremented by 0.01), which looks at what happens near 0.5.

TABLE 9.2

x	x^2	Change in x^2
0.46	0.2116	0.0091
0.47	0.2209	0.0093
0.48	0.2304	0.0095
0.49	0.2401	0.0097
0.50	0.2500	0.0099
0.51	0.2601	0.0101
0.52	0.2704	0.0103
0.53	0.2809	0.0105
0.54	0.2916	0.0107
0.55	0.3025	0.0109

Thus, the mapping diagram and numerical evidence have led us to a guess about squaring numbers. Before proceeding further with an algebraic analysis, we have to state exactly what this conjecture is. First of all, it is about the distance between numbers, which we can represent as $|b - a|$. We want to find out when the distance between the squares of two numbers is less than the distance between the numbers themselves, so suppose x_1 and x_2 are the numbers and $\left|x_2^2 - x_1^2\right| < \left|x_2 - x_1\right|$.

Then we would have

$$\left|x_2^2 - x_1^2\right| = \left|x_2 - x_1\right|\left|x_2 + x_1\right| < \left|x_2 - x_1\right|.$$

Dividing by $|x_2 - x_1|$ gives $|x_2 + x_1| < 1$. This is new information, which was not apparent from either the graphical or the numerical viewpoints. What does it mean, then, for a function to be a contraction *on an interval?*

DEFINITION: *f is a contraction on an interval I if $|f(x_2) - f(x_1)| < |x_2 - x_1|$ for all $x_1 \neq x_2$ in I.*

Thus, squaring is a contraction on an interval if $|x_1 + x_2| < 1$ for all x_1, x_2 in the interval, even for x_1 and x_2 arbitrarily close to each other. When x_1 and x_2 are very close, our condition becomes $|x_1 + x_2| \approx |2x_2| < 1$, or $|x_2| < 0.5$. This condition is both necessary and sufficient, and the following can be proved:

THEOREM: *Squaring is a contraction exactly on the interval* (–0.5, 0.5).

Since we've now opened up the bag of negative numbers, let's take a look at the squaring function on negatives, in this case [–2, 0]. (See fig. 9.7.) What are the new features here? Students will observe that mapping lines are now intersecting each other. What causes this? Answer: On this interval, we are dealing with a *decreasing* function, and we have already observed that for decreasing functions, mapping lines must cross.

Now let's put together the two mapping diagrams we have drawn for the squaring function. (See fig. 9.8.)

We see from these diagrams and our analysis that the squaring function moves the numbers inside the interval [–1/2, 1/2] *closer together*, whereas it moves numbers outside this interval farther apart.

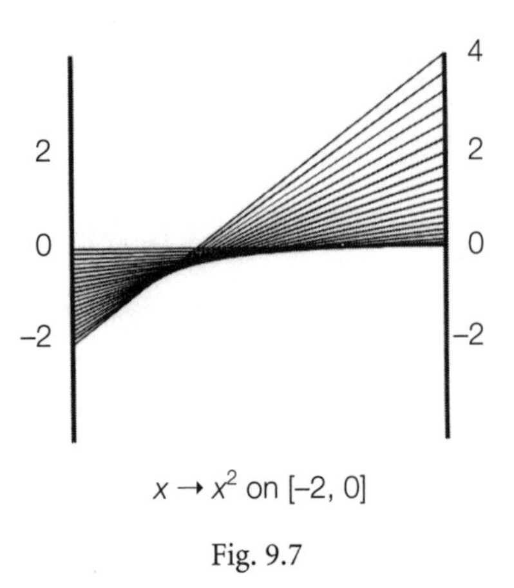

$x \to x^2$ on [–2, 0]

Fig. 9.7

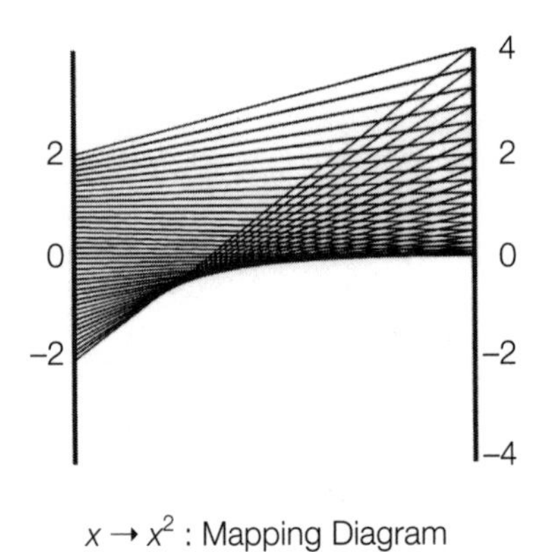

$x \to x^2$: Mapping Diagram

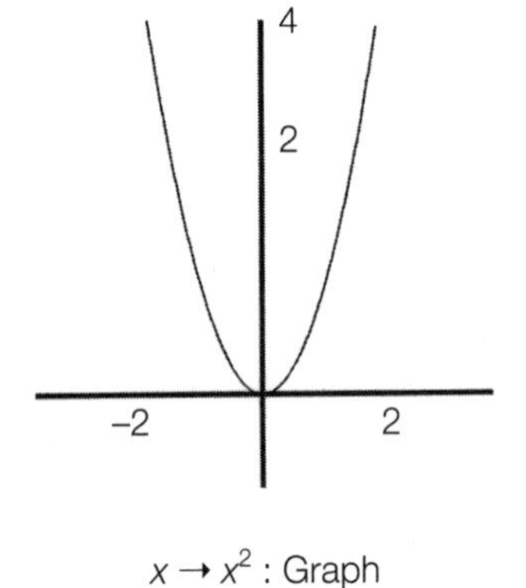

$x \to x^2$: Graph

Fig. 9.8

SCALINGS AND COVERINGS

One of the features of the squaring function that the mapping diagram shows more clearly than the graph is that the numbers in the interval [0, 4] all get "hit" twice by the numbers in [–2, 2] (0 is the image of only 0). This is sometimes expressed by saying that squaring is a *twofold covering* of [0, 4]. This property of the function shows up in the mapping diagram no matter what the scaling of the domain and range axes is. Until now, in order to investigate contractions and expansions, we have taken the scales on both axes to be the same. If we are interested in properties of functions that don't depend on relative *sizes,* such as being many-one rather than one-one, we can use whatever scales on the axes that are most convenient for our purposes. For example, figure 9.9 displays another view of the squaring function.

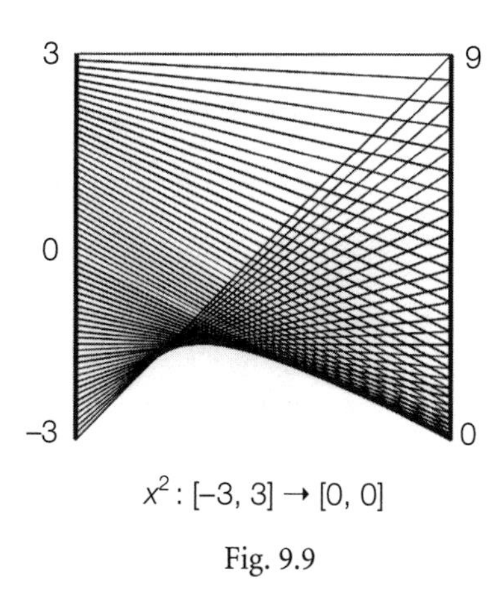

x^2 : [–3, 3] → [0, 0]

Fig. 9.9

Let's look at a more complicated function, this time a cubic: $f(x) = 2x^3 - 3x^2 - 12x$. As you can see from the diagrams in figure 9.10, f: [–3, 4] → [–45, 32].

We observe that this function is one-to-one in some places and many-to-one in others. Both the graph and the mapping diagram suggest that the numbers 7 and –20 in the range get hit twice, whereas all the numbers in the interval (–20, 7) get hit three times. $y = -20$ is the image of $x = 2$ and $x = -5/2$, whereas $y = 7$ is the image of $x = -1$ and $x = 7/2$. This is because the equations $f(x) = -20$ and $f(x) = 7$ have double roots at 2 and –1, respectively. Numbers outside [–20, 7] get hit only once. Having the students see these facts from the graphs and then verify them numerically and algebraically is what the "rule of three" was meant to encourage.

Of course, no discussion of many-one functions would be complete

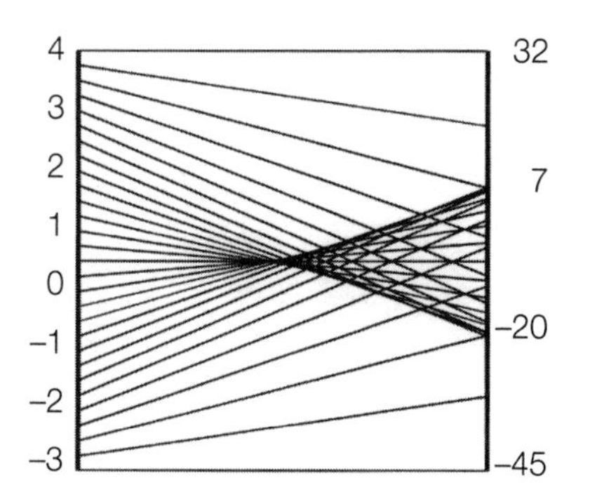

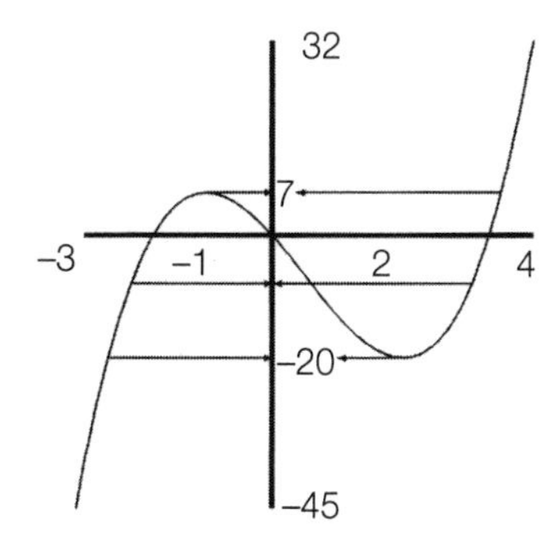

Fig. 9.10. Mapping diagram and graph for $x \rightarrow 2x^3 - 3x^2 - 12x$

without looking at trigonometric functions, which over a large domain are "very many to one." Figure 9.11 shows the cosine on the interval $[-2\pi, 2\pi]$.

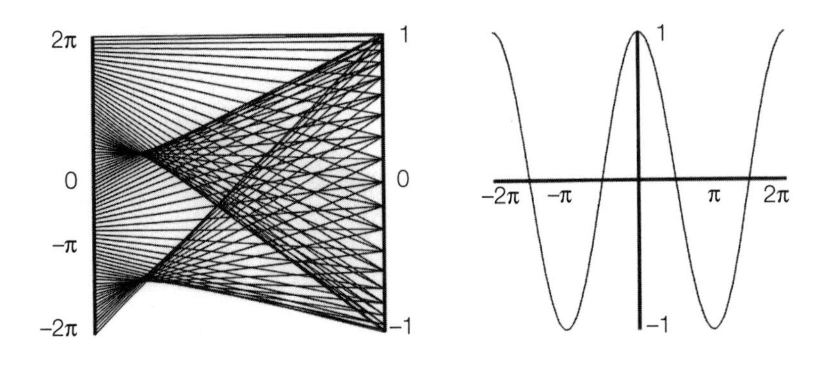

Fig. 9.11. Mapping diagram and graph for $x \rightarrow \cos(x)$

OTHER APPLICATIONS

Mapping diagrams are very useful in presenting inverse functions. For example, consider the function $x \rightarrow 3x + 2/5$ on the interval $[-4, 4]$. In the mapping diagram below (see fig. 9.12), we see that the image of $[-4, 4]$ contains $[-4, 4]$ but goes outside it as well.

We can ask our students, "What (sub)interval of $[-4, 4]$ gets mapped exactly onto $[-4, 4]$?" This leads naturally to asking which x gets mapped into -4 and which into 4 and then to the general question of "going backwards" in the mapping diagram, from points in the range to points in the domain. This, in turn, leads to the inverse mapping diagram for f^{-1}, the inverse of f, which is simply the mirror image of the diagram for f. (See fig. 9.13.)

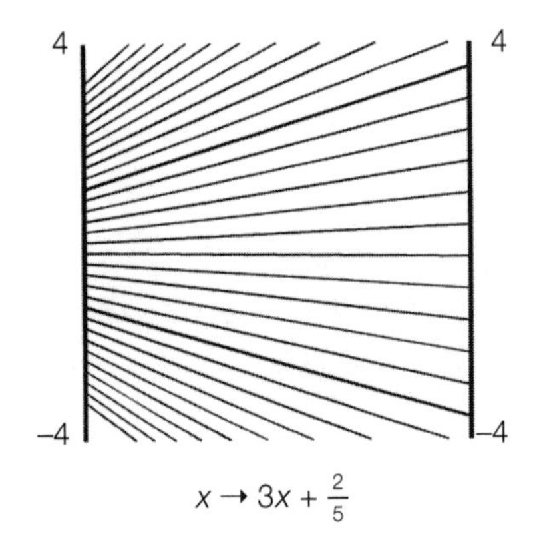

Fig. 9.12

Mapping diagrams are also eminently suited for visualizing compositions of functions. As an example, suppose $f(x) = 3x + 2/5$ and $g(x) = -1/2x + 3$. These functions involve stretchings and shiftings, and the graphs of f, g, and $g \circ f$ are not obviously related. Figure 9.14 shows the juxtaposed mapping diagrams for the composition $g(f(x))$.

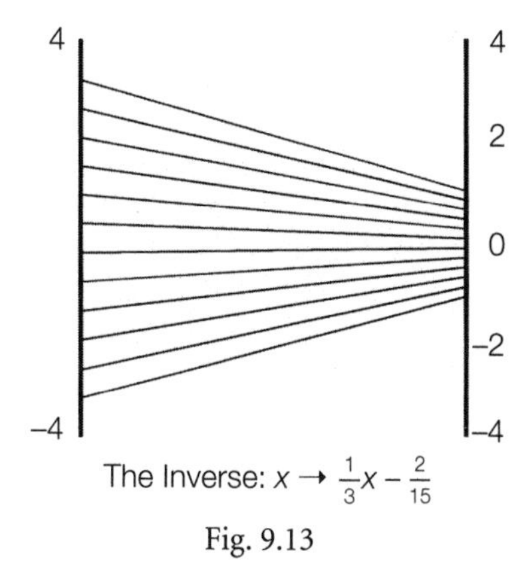

Fig. 9.13

Note that the first map is an expansion by a factor of 3, and the second map is a contraction by a factor of 1/2. This means that the first map carries each unit into three units, and each of these three units goes to 1/2 of a unit by way of the second.The effect of the composition is to send each unit into 3(1/2) units. Thus, for linear maps, scaling factors *multiply*. Since differentiable functions are locally linear, this helps explain the form of the chain rule in calculus.

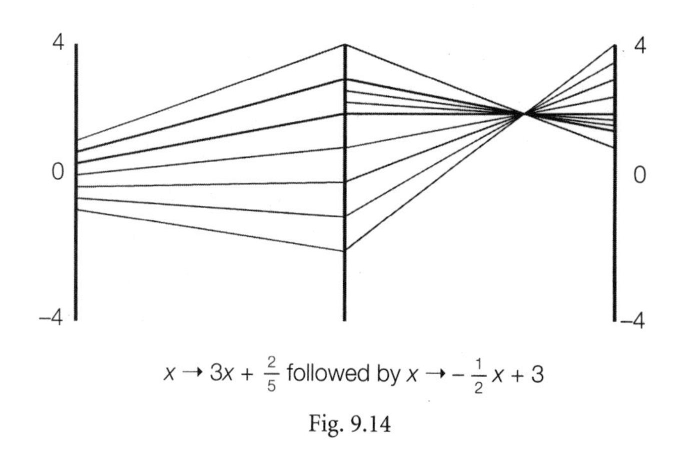

$x \rightarrow 3x + \frac{2}{5}$ followed by $x \rightarrow -\frac{1}{2}x + 3$

Fig. 9.14

A CLASSROOM EXPERIENCE

Some algebra 2 students were having a difficult time describing the behavior of functions of the form

$$f(x) = \frac{1}{x - a}.$$

The graphs they saw on their calculators made no sense to them, so their instructor suggested that they try mapping diagrams. They did one for the end behavior and one for the discontinuity. For the end behavior, they drew the domain and range lines on the board, at first with both going from –10 to 10, and then drew some mapping lines, staying away from $x = 2$. The students realized they needed to redraw the *range* line to see better what was happening near 0, so they changed scales. It became clear to them very quickly that $f(x) \to 0$ as $|x| \to +\infty$. (See fig. 9.15.)

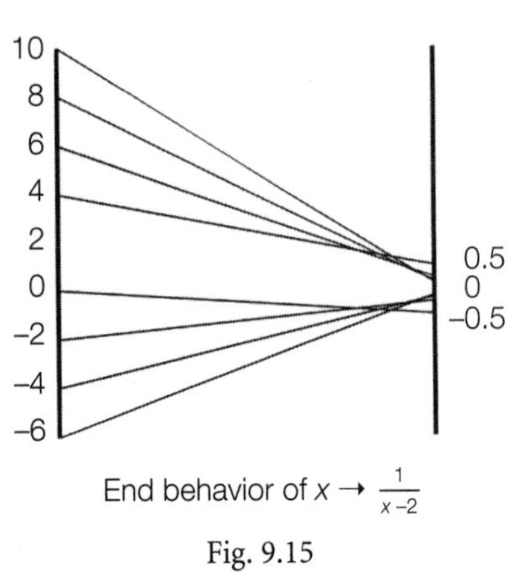

Fig. 9.15

Teachers are used to relating asymptotes on graphs to the limiting behavior of functions, but students don't always see things the way experts do. Here is an instance where mapping diagrams helped.

The class drew a second mapping diagram to see what the behavior of the same function is like near $x = 2$. The students began with $1 \le x \le 3$ and range [–10, 10] and soon realized they needed to look at points closer to 2. So they narrowed the domain to $1.5 \le x \le 2.5$ and drew mapping lines at points approaching $x = 2$ from both above and below. It soon became clear that $f(x) \to +\infty$ as $x \to 2^+$ and $f(x) \to -\infty$ as $x \to 2^-$. (See fig. 9.16.)

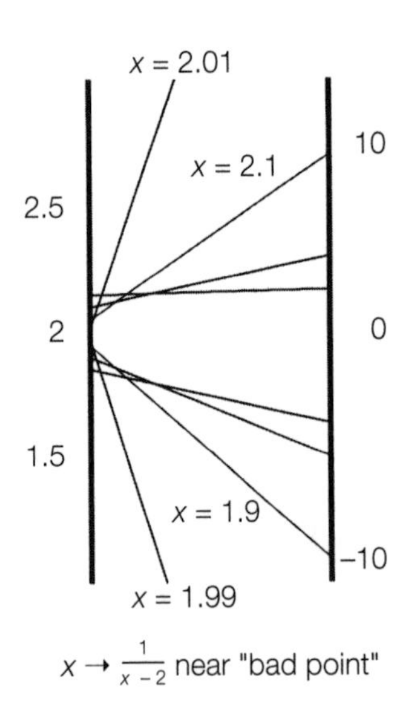

Fig. 9.16

This was the first time the class had used mapping diagrams. There was no assigned homework on them, yet when the class took the next exam, on rational expressions, there was a mapping diagram question that they all answered without any problem. It seems to be an easy and natural concept for them.

MORE ON DIAGRAMS, SOFTWARE, AND REFERENCES

The idea of mapping diagrams is not new. One of the earliest accounts is in Richmond (1963). Spivak (1967) uses a version of mapping diagrams in his wonderful calculus book, and Brieske (1973, 1978) has a short description of them in his 1973 article and a discussion of their relations to calculus in his later paper. Goldenberg (1992) and others also discuss a form of mapping diagrams and their relation to the development of the concept of function.

A more complete account can be found in Bridger (1996). This article introduces some of the rather elegant theory surrounding mapping diagrams, including focal points, fixed points, and enveloping curves. It also describes the Function Visualizer, a piece of public domain software by the author, which draws both Cartesian graphs and mapping diagrams and uses animation to show how functions distribute the images of their domain points over their range. This software is available, for Macs and PCs, from Mark Bridger's home page, www.math.neu.edu/~bridger/mathindex.html, under "Mapping Diagrams Material."

A number of teachers in the United States, including the authors, have been experimenting with mapping diagrams and Visualizer software in the classroom. Students welcome the ideas, since they are new, straightforward, and make the concept of function easier to understand. The software has also been tested in Israel with similar results. The authors are now in the process of designing workbook materials for independent and classroom use.

REFERENCES

Boyer, Carl B. *A History of Mathematics.* 2nd ed., revised by Uta C. Merzbach. New York: John Wiley & Sons, 1989.

Bridger, Mark. "Dynamic Function Visualization." *College Mathematics Journal* 27 (November 1996): 361–69.

Brieske, Thomas. "Functions, Mappings, and Mapping Diagrams." *Mathematics Teacher* 66 (May 1973): 463–68.

———. "Mapping Diagrams, Continuous Functions, and Derivatives." *College Mathematics Journal* 9 (1978): 67–72. (Reprinted in *A Century of Calculus, Part II,* edited by T. M. Apostol, D. H. Mugler, D. R. Scott, A. Sterrett, Jr., and A. E. Watkins, pp. 113–18. Washington, D.C.: Mathematical Association of America, 1992.)

Calinger, Ronald, ed. *Classics of Mathematics.* Englewood Cliffs, N.J.: Prentice Hall, 1995.

Goldenberg, Paul, Philip Lewis, and James O'Keefe. "Representation and the Development of a Process Understanding of Function." In *The Concept of Function: Aspects of Epistemology and Pedagogy,* edited by Ed Dubinsky and Guershon Harel, pp. 235–60. Washington, D.C.: Mathematical Association of America, 1992.

Katz, Victor. *A History of Mathematics.* 2nd ed. Reading, Mass.: Addison-Wesley, 1998.

Richmond, Donald E. "Calculus—a New Look." *American Mathematical Monthly* 70 (1963): 415–23. (Reprinted in *A Century of Calculus, Part I,* edited by T. M. Apostol, H. E. Chrestenson, C. S. Ogilvy, D. E. Richmond, and

N. J. Schoonmaker, pp. 84–92. Washington, D.C.: Mathematical Association of America, 1992.

Spivak, Michael. *Calculus.* 3rd ed. Houston, Tex.: Publish or Perish, 1994.

APPENDIX

We sketch, using analytic geometry, the following proof that linear functions have focal points. Set up a coordinate system with a horizontal t-axis and a vertical s-axis. Represent the domain axis in the mapping diagram by the line $t = 0$ and the range by the line $t = 1$. Then show that a mapping line connects $(0, x)$ with $(1, f(x))$, so its slope is $f(x) - x$ and its s-intercept is x; thus, its equation (as a line) is $s = (f(x) - x)t + x$. Now suppose $f(x) = ax + b$ and rewrite this equation first with $x = x_1$ and then with $x = x_2$. Solving these two equations in the unknowns, s and t, we get the intersection point of the mapping lines: $P = (1/(1 - a), b/(1 - a))$. Since P is independent of x_1 and x_2, it lies on all mapping lines and, hence, is the focal point.

10

A Multirepresentational Journey through the Law of Cosines

Daniel Scher

E. Paul Goldenberg

THE standard law of cosines proof is a model of economy. With just the Pythagorean theorem and a definition of cosine, five steps (fig. 10.1) yield the result $c^2 = a^2 + b^2 - 2ab\cos(C)$.

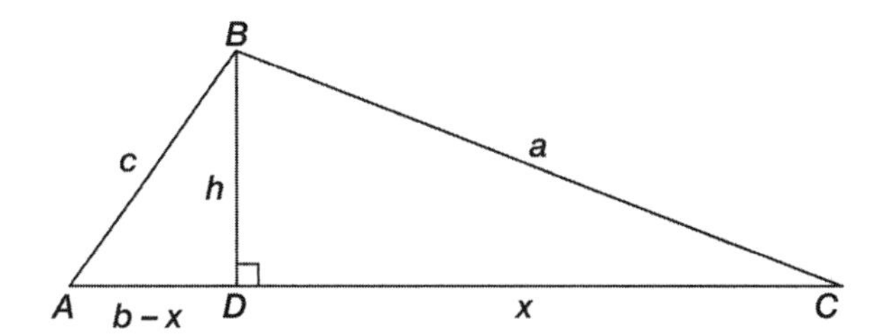

1. Applying the Pythagorean theorem to $\triangle ABD$: $(b - x)^2 + h^2 = c^2$
2. Applying the Pythagorean theorem to $\triangle CBD$: $x^2 + h^2 = a^2$, or $h^2 = a^2 - x^2$
3. Substituting for h^2 in equation 1: $(b - x)^2 + (a^2 - x^2) = c^2$
4. Simplifying 3 and reorganizing: $c^2 = a^2 + b^2 - 2bx$
5. Rewriting x as $a\cos(C)$: $c^2 = a^2 + b^2 - 2ab\cos(C)$

(When the perpendicular does not meet $\overline{AC}$, the proof still works with only minor variations.)

Fig. 10.1. An algebraic proof of the law of cosines

This work was supported by the National Science Foundation grant RED-9453864 as part of the Epistemology of Dynamic Geometry Project at Education Development Center, Inc. Thanks to David Dennis for introducing us to the geometric law of cosines approach presented here.

Likened to a walk, this derivation is less a leisurely stroll than a brisk hike. In figure 10.1's proof, the final result seems to "pop out" without advance warning. Is there a way of approaching the law of cosines that builds more on our intuition? In the Pythagorean theorem, a^2, b^2, and c^2 represent areas of squares. Is there a corresponding geometric interpretation of the $2ab\cos(C)$ term? These questions remain unanswered in the classic proof.

To gain a better sense of the mathematical scenery, we propose to catch our breath, slow down, and pursue some unexplored back roads leading to the same destination. Our paths will correspond to two alternative representations of the law of cosines—numerical and geometric. By changing the landscape of our proof, we will discover that our representations yield intriguing insights—both mathematical and pedagogical—into the underpinnings of this theorem.

The methods below originate in our *Connected Geometry* curriculum (Education Development Center 2000) and benefit from interactive geometry software such as the Geometer's Sketchpad (Jackiw 1995). You'll find downloadable files containing the Sketchpad models described in this article at www.addr.com/~dscher/cosines.html. The Web site also contains interactive Java applets of these same experiments that can be explored online with just your Web browser.

A NUMERICAL REPRESENTATION

In figure 10.1, the symbol x represents DC. The segment could be described by its length, 3 inches, but a single piece of numerical data typically tells little. Although no amount of measuring can ever substitute for a proof, it is intriguing to ask what role numerical data can play when they are available in vast quantities that can be easily manipulated and analyzed.

Interactive geometry software, with its measurement and calculation features, provides this very functionality. The value of any geometric length or calculation updates itself continuously as one drags and deforms a construction. By representing a malleable geometric model through its numerical features, students can search for patterns, devise experiments to test conjectures, and use their knowledge of algebra to make sense of collected data.

The following numerical approach to the law of cosines illustrates well this mixture of experimentation and deduction.

The Construction

With interactive geometry software, draw an arbitrary triangle ADE, place a moveable point B on $\overline{AD}$, and construct $\overline{BC}$ parallel to $\overline{DE}$ (see fig. 10.2, the first downloadable sketch, or the prebuilt Java model online).

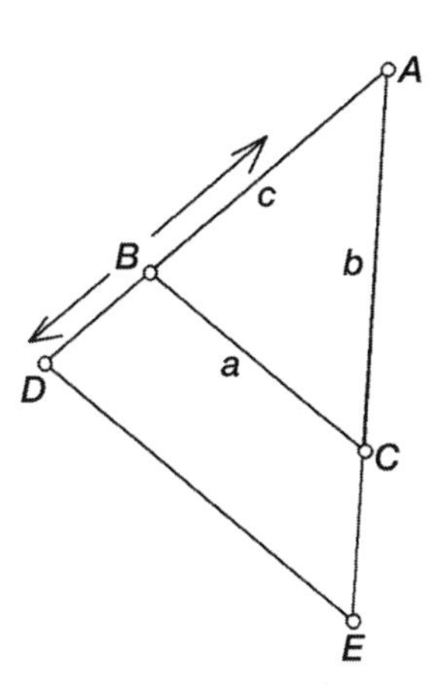

Fig. 10.2. Constructing similar triangles, with point B moveable

As B slides along $\overline{AD}$, $\overline{BC}$ moves with it, remaining parallel to $\overline{DE}$. Thus $\triangle ABC$ may be viewed as a family of similar triangles all similar to $\triangle ADE$ and sharing invariant angle measures. Of course, angles are not the only invariants worth noticing. Some experimentation and computation with the software leads to the observation that whereas the values of b/a, c/a, and c/b depend on the shape of $\triangle ABC$, these ratios remain fixed for all locations of B.

Students skilled in looking for numerical invariants will know that arithmetic combinations of invariants are themselves invariant, and they may propose "compounds" like $a/b + b/a$ or $c/a \cdot c/b$ or even (utterly implausibly, but correctly) $1/2 \cdot (a/b + b/a - c/a \cdot c/b)$.

In mathematical investigations, one often gains insights by examining special cases. When $\angle C$ measures 90°, the invariant value of $a^2 + b^2 - c^2$ is zero: this is the Pythagorean theorem. Students can check by adjusting their construction so that $m\angle C = 90°$ (see fig. 10.3). Sliding B along $\overline{AD}$ generates a continuum of right triangles ABC, in all of which $a^2 + b^2 - c^2$ has the expected invariant value of zero (or nearly so, depending on computer round-off error). The teacher can now seed the investigation with a new (very plausible, but incorrect) conjecture. When $m\angle C \neq 90°$, might $a^2 + b^2 - c^2$ again remain constant, but at some value k other than zero?

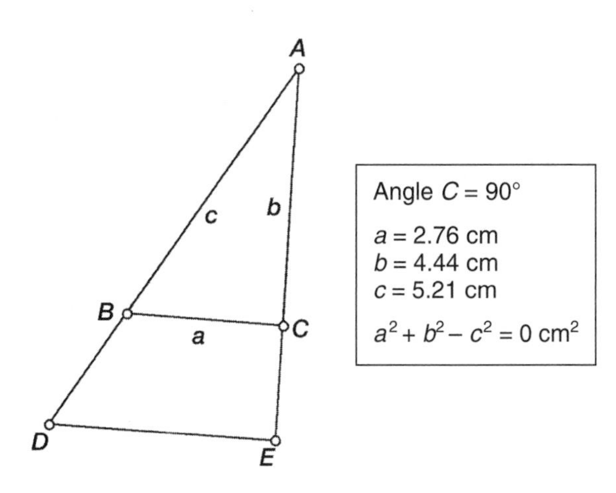

Fig.10. 3. The Pythagorean theorem generates an invariant.

Experimenting shows that it does not. Table 10.1 displays the situation when $m\angle C = 72°$. The table contains four sets of data—the values of a^2, b^2, c^2, and the conjectured invariant $a^2 + b^2 - c^2$ for four locations of B along $\overline{AD}$. As the size of $\triangle ABC$ increases, so does the value of $a^2 + b^2 - c^2$. The conjecture fails, but perhaps it can be salvaged.

TABLE 10.1
Data Sets for Four Locations of Point B along $\overline{AD}$

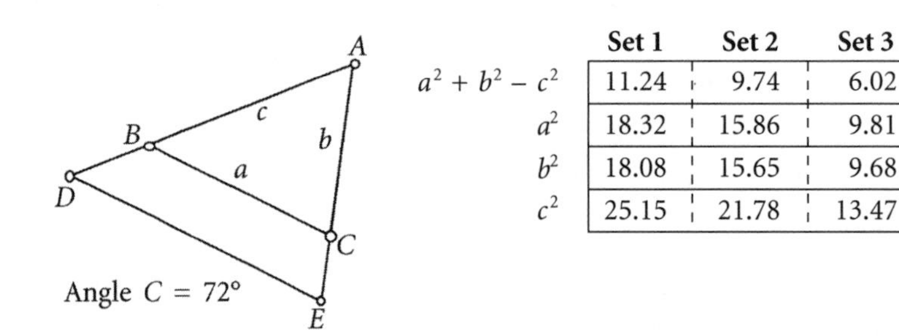

	Set 1	**Set 2**	**Set 3**	**Set 4**
$a^2 + b^2 - c^2$	11.24	9.74	6.02	4.32
a^2	18.32	15.86	9.81	7.04
b^2	18.08	15.65	9.68	6.94
c^2	25.15	21.78	13.47	9.66

In fact, a hint resides right in the statement of failure: as *this* increases, so does *that*. When two values increase and decrease in unison, it's possible that their ratio or difference is constant. To calculate a ratio here, we need a way to represent the size of $\triangle ABC$. For simplicity's sake, a suitable size measure is $ab/2$, or just ab. (Of course

$$\text{Area } (\triangle ABC) = \frac{ab \cdot \sin(C)}{2},$$

but $ab/2$ is a simple approximation that continues our right-triangle analogy. If the triangle had a right angle at C, its area would be exactly $ab/2$.). Is

$$\frac{a^2 + b^2 - c^2}{ab}$$

constant? Experimentation indicates, and a little algebra proves, the following:

$$\frac{a^2 + b^2 - c^2}{ab} = \frac{a}{b} + \frac{b}{a} - \frac{c^2}{ab} = \frac{a}{b} + \frac{b}{a} - \frac{c}{a} \cdot \frac{c}{b}$$

The data in table 10.1 suggest other possible constant ratios, too. As $a^2 + b^2 - c^2$ increases, so do the values of the square terms. Might ratios like

$$\frac{a^2 + b^2 - c^2}{a^2}$$

be invariant for similar triangles? Algebra again provides confirmation:

$$\frac{a^2 + b^2 - c^2}{a^2} = 1 + \left(\frac{b}{a}\right)^2 - \left(\frac{c}{a}\right)^2$$

Following the same algebraic lead yields a total of six invariant ratios (see table 10.2).

TABLE 10.2

Six Computations That Are Invariant When the Size of △ABC Varies but Its Shape Remains Fixed

$\frac{a^2 + b^2 - c^2}{a^2}$	$\frac{a^2 + b^2 - c^2}{b^2}$	$\frac{a^2 + b^2 - c^2}{c^2}$
$\frac{a^2 + b^2 - c^2}{ab}$	$\frac{a^2 + b^2 - c^2}{ac}$	$\frac{a^2 + b^2 - c^2}{bc}$

Modifying the Experiment

Although these six ratios grew naturally enough out of the dynamic experiment, are any of them more "special" than the others? One way to check is to test them under different conditions. How do they behave, for instance, with the similarity restriction relaxed? Using the software or the online Java applet to compute the six values in table 10.2, students can drag points D, A, and E around the screen to vary the shape (angle measurements) of $\triangle ABC$. Anywhere from two to six pairs of students' eagle eyes can monitor the range of values assumed by the six quantities. Table 10.3 shows several sets of the data, each for a different $\triangle ABC$.

At first blush, it seems that when $\triangle ABC$ changes shape, none of the six computations remains invariant. Yet closer inspection of table 10.3 reveals that one ratio is more "tame" than the others. Five of them vary between seemingly arbitrary limiting values (some become infinite; others have finite but unrecognizable limits). One of them,

$$\frac{a^2 + b^2 - c^2}{ab},$$

varies neatly between –2 and 2! Interestingly, this ratio already has a special distinction, being a natural encoding of our original observation that the value of $a^2 + b^2 - c^2$ varies with the size of the triangle.

This new experiment drew attention to one computation but was really too broad. Whereas the first experiment restricted the shape of $\triangle ABC$ letting only the size vary, the second experiment permitted *too much* to vary. Let's adopt a compromise. We will allow the shape of $\triangle ABC$ to vary but keep the measure of $\angle C$ fixed. To test this setup, open the second downloadable sketch or follow along online.

With this refined experiment, exactly one of the six computations remains invariant:

$$\frac{a^2 + b^2 - c^2}{ab}$$

For the student who has no experience with trigonometric relationships, this is already an interesting result: a function of $\angle C$ that has a fixed value—call it k—for all changes in the size or shape of $\triangle ABC$ that keep $\angle C$ fixed but that varies between –2 and 2 with changes in $\angle C$. And because this salvaged a conjecture based on the Pythagorean theorem, it has created a new, generalized relation. The observation

$$\frac{a^2 + b^2 - c^2}{ab} = k$$

is easily rewritten as $a^2 + b^2 = c^2 + abk$.

TABLE 10.3
Data Sets Generated for Six Different Configurations of $\triangle ABC$

	Set 1	Set 2	Set 3	Set 4	Set 5	Set 6
$\frac{a^2 + b^2 - c^2}{a^2}$	1.60	16.78	1.81	0.19	–65.15	–0.30
$\frac{a^2 + b^2 - c^2}{b^2}$	2.51	0.24	2.21	13.06	–0.06	–13.43
$\frac{a^2 + b^2 - c^2}{c^2}$	38.77	0.30	194.35	0.23	–0.06	–0.22
$\frac{a^2 + b^2 - c^2}{a \cdot b}$	2.00	1.99	2.00	1.58	–1.98	–2.00
$\frac{a^2 + b^2 - c^2}{a \cdot c}$	7.86	2.26	18.76	0.21	–1.92	–0.26
$\frac{a^2 + b^2 - c^2}{b \cdot c}$	9.86	0.27	20.71	1.74	–0.06	–1.73

For students who know some basic trigonometry, the result is even more amazing. "Hmm," they might say, "what do we know that varies between –2 and 2 and depends only on angle measurement and not side lengths? Nothing. But if it varied from –1 to 1.... Aha! Multiply

$$\frac{a^2 + b^2 - c^2}{ab}$$

by one-half to obtain

$$\frac{a^2 + b^2 - c^2}{2ab}.$$

This *appears* to behave something like the sine or cosine. Could it be one of them?"

Yes. Take the law of cosines, $c^2 = a^2 + b^2 - 2ab\cos(C)$, and solve for $\cos(C)$. The result is

$$\cos(C) = \frac{a^2 + b^2 - c^2}{2ab}.$$

By the way, after all that excitement, it is sad to report that we have failed ... sort of. This is *not* a proof. The invariance of

$$\frac{a^2 + b^2 - c^2}{2ab}$$

is secure enough when a, b, and c vary in ways that preserve the ratios of any two of them, but the looser restriction that only $\angle C$ be fixed does not guarantee invariant ratios. In fact, what we have is a "derivation without proof." The interaction between experimentation and solid mathematical or algebraic reasoning generates a computation appearing to have many remarkable and desirable properties. Because the computation also happens to fit a known formula—the law of cosines—it sheds new light on it.

A Geometric Representation

Our second approach provides a concrete geometric representation for each term in the law of cosines expression (Dintzl 1931). We start with a familiar friend: figure 10.4 shows a square with side c subdivided into two smaller squares and two congruent rectangles. A look at their areas tells us that $c^2 = (a + b)^2 = a^2 + 2ab + b^2$.

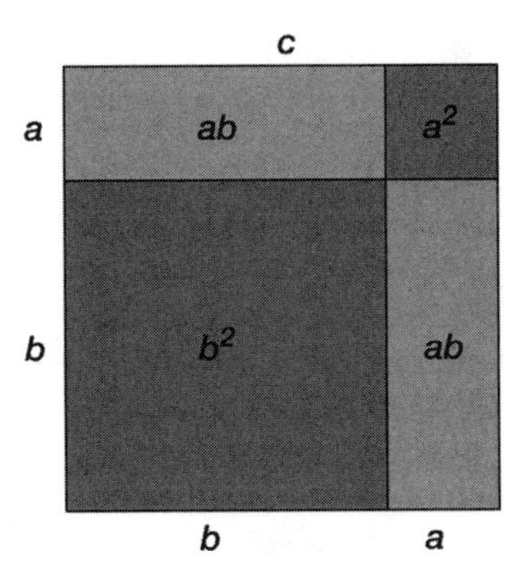

Fig. 10.4. A geometric interpretation of $c^2 = (a + b)^2 = a^2 + 2ab + b^2$

Imagine a flexible physical model of figure 10.4 in which the common vertex of the two internal squares, a^2 and b^2, can be moved. Two restrictions apply: the vertices of the parts remain connected as they are (with the appropriate ones remaining attached to the four vertices of the outside square), and all three squares remain square (with the two internal squares rotating and growing or shrinking). Figure 10.5 shows how the two rectangles squish into parallelograms as their common vertex *G* moves. What geometric and algebraic relations reside here?

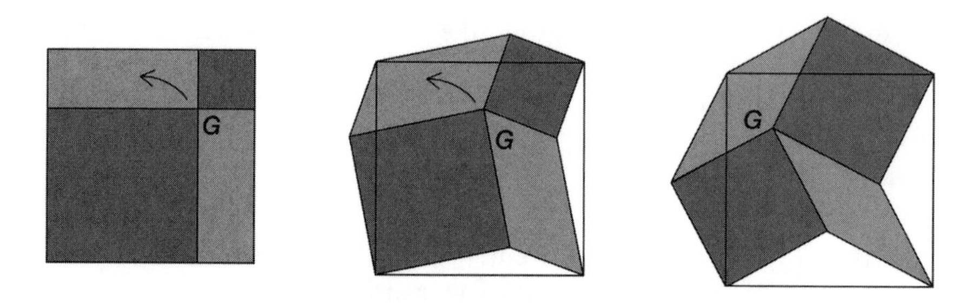

Fig. 10.5. Generalizing the construction from figure 10.4

You can develop a sense for this model by building one with dynamic geometry software. Figure 10.6 describes one such construction method. You'll also find a complete working model by opening the third downloadable sketch or viewing the online Java applet.

With the construction complete, you can drag point *C* (keeping $\angle C$ obtuse) and observe how the model reacts. Certain relations appear to hold regardless of *C*'s location:

- Parallelograms *BCGH* and *FDIG* are congruent with sides of length *a* and *b*.
- Triangles *ABC, BEH, DEI,* and *ADF* are all congruent.

Proving these statements makes a nice exercise.

In figure 10.4, the two small squares and two rectangles fit snugly into the larger square. Does a similar observation apply to this new model? Imagine translating $\triangle DEI$ down to $\triangle ABC$ and $\triangle ADF$ across to $\triangle BEH$ in figure 10.6. Now the shaded regions—what began as two squares and two parallelograms—fit perfectly into *ADEB.* Viewed differently, figure 10.7 shows a tessellation in which copies of square *ADEB* are superimposed on a repeating pattern of the two small squares and parallelograms. Since both tessellations tile the plane, they cover the same area. So at least while $\angle C$ remains obtuse, Area(*ADEB*) = Area(*GIEH*) + Area(*AFGC*) + 2Area(*BCGH*) or equivalently, $c^2 = a^2 + b^2 + 2\text{Area}(BCGH)$.

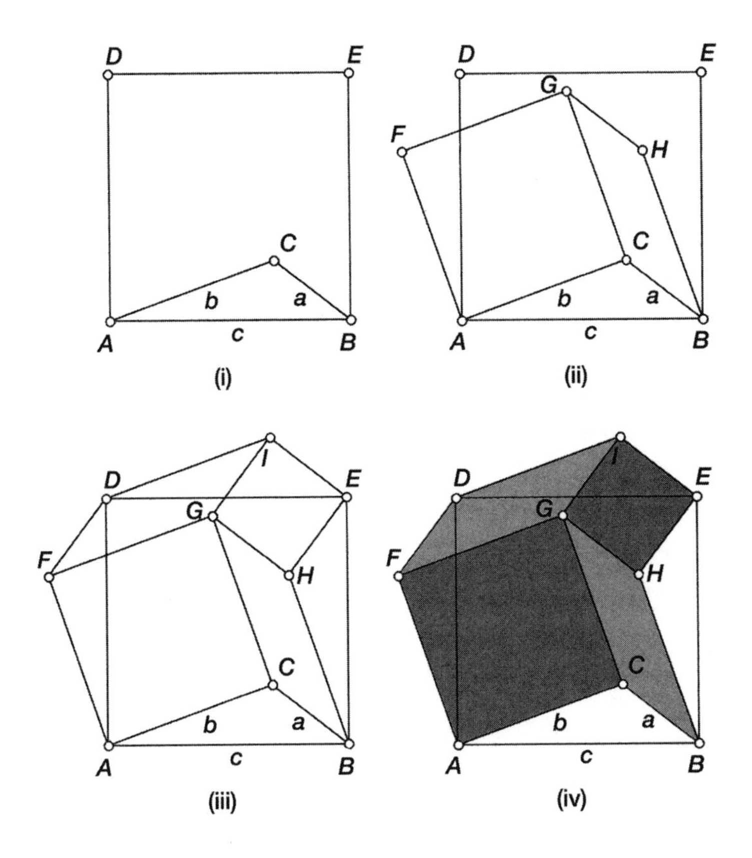

i. Draw an arbitrary triangle ABC with $\angle C$ obtuse.
   Construct a square with side $\overline{AB}$ that covers the triangle.

ii. Construct a square on side $\overline{AC}$ of the triangle.
   Construct a parallelogram with sides $\overline{CB}$ and $\overline{CG}$.

iii. Construct a square with side $\overline{GH}$.
   Construct a parallelogram with sides $\overline{FG}$ and $\overline{GI}$.

iv. Shade the interiors of the two squares and two parallelograms.

Fig.10. 6. Construction steps for building a model of figure 10.5

Notice the resemblance of the statements above to the Pythagorean theorem. For right triangles, the area of a large square is equal to the sum of the areas of two smaller squares. For obtuse triangles, we need only add the areas of two congruent parallelograms to maintain the equality.

Fig. 10.7. A tessellation derived from the construction in figure 10.6

Students can now apply the formula above concretely to determine an unknown side of an obtuse triangle. Figure 10.8 shows $\triangle ABC$ with $AC = 6$ and $BC = 4$. If the height of parallelogram $BCGH$ is 2, what is the length of $\overline{AD}$? The area relationship above says

$$AB^2 = 4^2 + 6^2 + 2(6 \cdot 2),$$

so $AB = \sqrt{76}$.

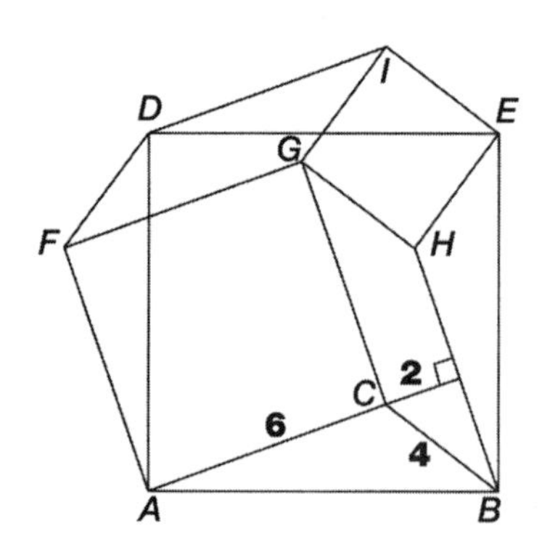

Fig. 10.8. What is the length of $\overline{AB}$?

Without using any trigonometry, students have just solved a question using the law of cosines! Puzzled? Look again at the statement $c^2 = a^2 + b^2 + 2\text{Area}(BCGH)$. It is certainly close to the cosines formula, but where is the cosine term?

Figure 10.9's labeling helps solve the mystery. With $\angle ACB$ labeled as θ, $m\angle CBJ = \theta - 90$. The area of $BCGH$ is $BH \cdot CJ$, but $BH = CG = AC = b$, and $CJ = a \sin(\theta - 90) = -a \cos \theta$. Thus,

$$\begin{aligned}\text{Area}(BCGH) &= BH \cdot CJ \\ &= -ab \cos \theta.\end{aligned}$$

The formula $c^2 = a^2 + b^2 + 2\text{Area}(BCGH)$ now becomes the familiar $c^2 = a^2 + b^2 - 2ab \cos \theta$.

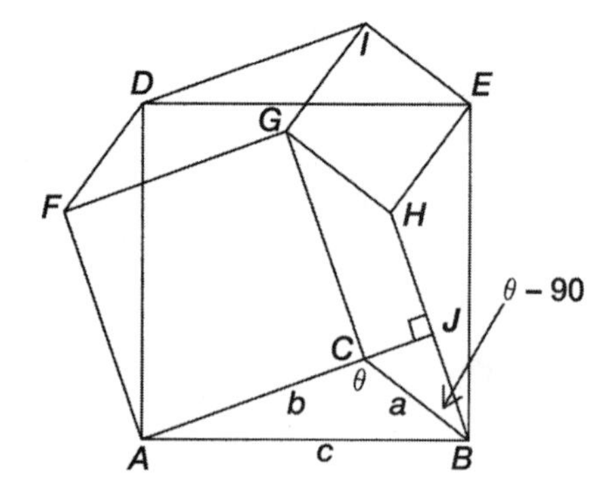

Fig. 10.9. Relating the construction to the law of cosines

Generalizing the Dissection Picture

Throughout the dissection investigation, we've been careful to keep $\angle C$ obtuse. The picture becomes slightly more complicated when $\angle C$ is acute (fig. 10.10). In this instance,

$$\text{Area}(ADEB) = \text{Area}(GIEH) + \text{Area}(AFGC) - \text{Area}(BCGH) - \text{Area}(FDIG).$$

To verify the statement above, use a pencil to shade the squares *GIEH* and *AFGC* (noting that some regions get shaded twice), and then subtract the parallelogram regions *BCGH* and *FDIG*. The remaining regions can be made to fit within *ADEB*. Working through the algebra of the situation again gives the law of cosines formula.

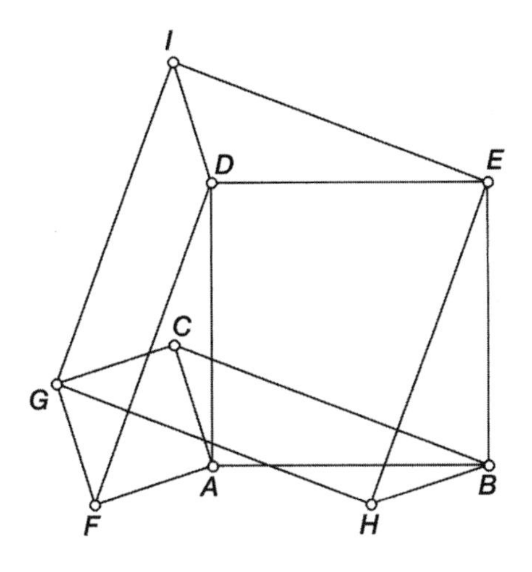

Fig. 10.10. The construction when $\angle C$ is acute

Finally, in figure 10.11 where the measure of $\angle C$ equals 90 degrees, the two parallelograms disappear. What is left is a geometric dissection proof of the Pythagorean theorem (slide $\triangle DEI$ downward and $\triangle ADF$ to the right, to fit into square *ABED*).

It is remarkable how this geometric representation links together several mathematical results through the fluid movement of its parts. Dragging point *C* causes angle *C* to vary from 180 degrees, to an obtuse angle, to 90 degrees, to an acute angle. As it does so, the geometric construction deforms in a continuous manner, linking established results ($(a + b)^2 = a^2 + 2ab + b^2$ and the Pythagorean theorem)

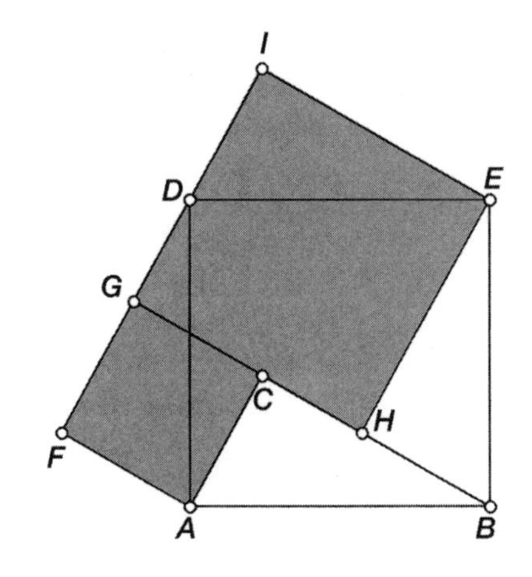

Fig. 10.11. Relating the construction to the Pythagorean theorem

with new generalizations (the law of cosines), all in a seamless progression (see fig. 10.12).

$(a+b)^2 = a^2 + 2ab + b^2 \rightarrow$ law of cosines (obtuse angles) $\rightarrow$
Pythagorean theorem $\rightarrow$ law of cosines (acute angles)

Fig. 10.12. The progression of theorems illustrated as $\angle C$ shrinks continuously from 180 degrees to an acute angle.

CONCLUSION

As companions to the traditional algebraic law of cosines proof in figure 10.1, the numerical and geometric representations discussed above provide connections extending well beyond the Pythagorean theorem and trigonometry. Reasoning about calculations, designing experiments, searching for invariants, deforming an established result in a continuous way, and analyzing dissections are just some of the mathematical topics and ways of thinking made possible by the diverse representations.

Although our article focused on the law of cosines, this theorem was certainly not unique in its adaptability to a multirepresentational approach. We offer as a next step the challenge of reconceptualizing other traditional proofs in ways that promote novel and diverse mathematical thinking.

REFERENCES

Dintzl, E. "Über die Zerlegungsbeweise des Verallgemeinerten Pythagoreischen Lehrsatzes." *Zeitschrift für Mathematischen und Naturwissenschaftlichen Unterricht* 62 (1931): 253–54.

Education Development Center. *Connected Geometry.* Chicago: Everyday Learning Corp., 2000.

Goldenberg, E. Paul. "An Inefficient Route to the Cosine Law." *International Journal of Computers for Mathematical Learning* 3 (1998): 185–93.

Jackiw, Nicholas. The Geometer's Sketchpad. Software. Berkeley, Calif.: Key Curriculum Press, 1995.

11

Representations of Reversal
An Exploration of Simpson's Paradox

Lawrence Mark Lesser

When updating its *Standards* documents, the National Council of Teachers of Mathematics (NCTM) added a pre-K–12 Standard on representation, urging that students be able to develop a repertoire of mathematical representations that can be used purposefully and flexibly to model and interpret physical, social, and mathematical phenomena (NCTM 2000). This article aims to explore the potential of including multiple representations in one's teaching repertoire through an accessible phenomenon for which full insight is not obvious from using only the single most common representation. The phenomenon chosen, Simpson's paradox, can be concisely defined as the reversal of a comparison when data are grouped. In this particular example, we will see that it is possible for women to be hired at a higher rate than men within each of two departments but at a *lower* rate than men when the data from both departments are pooled together.

The Relevance of Simpson's Paradox

Simpson's paradox was first noted in 1951 by the British statistician E. H. Simpson but was discussed as early as 1903 by the Scottish statistician George Yule (Wagner 1983). Simpson's paradox can involve a comparison of overall rates, ratios, percentages, proportions, probabilities, averages, or measurements that are weighted averages of subgroup counterparts. Students are likely vulnerable to this paradox if they have the related "averaging the averages" misconception, in which they compute the ordinary average in problems requiring the weighted average. In a weighted average, an overall average is computed by weighting the individual averages by the sizes of their corresponding individual groups. For example, if the average final exam

The author expresses appreciation to Ralph Cain, Lindsey Eck, the 2001 Yearbook editorial panel, and his secondary school methods students.

score in a 30-student class is 100 and the average final exam score in a 10-student class is 60, the overall average for all 40 students is not the "unweighted" mean $(100 + 60)/2 = 80$ but is obtained as a weighted mean: $((100)(30) + (60)(10))/40 = 90$. When some courses are worth more credit hours than others, a student's overall grade-point average is a weighted average as well. A geometrical interpretation of the weighted mean is described by Hoehn (1984) and will be discussed later.

Simpson's paradox was chosen for the investigation for several reasons:

1. It is simple enough to encounter (e.g., table 11.1) with mere fraction arithmetic and yet complex enough to model with tools spanning a broad range of high school mathematics content, generating many different representations.

2. A paradox can motivate students (Movshovitz-Hadar and Hadass 1990; Wilensky 1995; Lesser 1998).

3. Its structure relates to common student misconceptions regarding weighted means or even the addition of fractions. Noting that students are taught (correctly) that the statements

$$a/b > e/f \text{ and } c/d > g/h \text{ imply } a/b + c/d > e/f + g/h,$$

Mitchem (1989) suggests that students who (incorrectly) add fractions by adding the numerators and adding the denominators would assume that $(a + c) / (b + d)$ always exceeds $(e + g) / (f + h)$ and thus be vulnerable to the paradox.

4. It provides many opportunities to explore "both the mathematical and developmental advantages and disadvantages in making selections among the various models" (NCTM 1991, p. 151).

5. It allows "a view of a real-world phenomenon ... through an analytic structure imposed on it" (NCTM 2000, p. 70).

This phenomenon is not contrived: it has actually occurred in many natural situations, including university admission rates (male versus female), fertility rates (rural versus urban), death rates (young versus old), death penalty cases (black versus white), categories of federal tax rates, and various baseball statistics (Bassett 1994; Bickel, Hammel, and O'Connell 1975; Cohen 1986; Moore and McCabe 1993; Wagner 1982a). Simpson's paradox underscores the pitfall of basing a conclusion on only a single average and demonstrates a general need for intuition to be checked against mathematical arguments. Exploring Simpson's paradox may also stimulate greater awareness of what one is averaging over, such as the phenomenon that a university's mean class size averaged over students is never smaller than the mean class size averaged over classes (Hemenway 1982; Movshovitz-Hadar and Webb 1998).

REPRESENTATIONS OF SIMPSON'S PARADOX

What are some ways to think about Simpson's paradox? This article presents many ways and invites the reader to grab pencil and paper and explore them. The reader is also encouraged to reflect on the "relative strengths and weaknesses of various representations for different purposes" (NCTM 2000, p. 70). Which ones seem "new" or applicable to other mathematics content?

Numerical or Tabular Representations

This section begins with a numerical or tabular representation because it is concrete and is the most common representation that textbooks use for Simpson's paradox. Table 11.1 is a $2 \times 2 \times 2$ table involving the three categorical variables gender (male or female), department (social sciences or physical sciences), and employment application status (hired or denied). The numbers were chosen for ease of computation and to draw attention to the role of where the larger and smaller cell sizes were located. (Later in the article, table 11.2 offers a similarly behaving data set that is more subtle in appearance.) It is routine to verify that within each department, women are hired at a higher rate than men (since $30/80 = .375 > .25 = 5/20$ and $15/20 = .75 > .625 = 50/80$), yet are hired at a lower rate than men for the overall situation: $(30 + 15)/100 = .45 < .55 = (5 + 50)/100$.

TABLE 11.1
Hiring Data (by Gender and Department)

	Social Sciences		Physical Science		Overall	
	Male	Female	Male	Female	Male	Female
Hired	5	30	50	15	55	45
Denied	15	50	30	5	45	55
Total applied	20	80	80	20	100	100

As Wagner (1982b) states, "Because this situation occurs at the level of a purely descriptive data analysis, it can easily bewilder the statistically naïve observer" (p. 47). Indeed, many students have responded with the reaction "I follow the arithmetic, but I still don't believe the result." The numerical representation is undeniably effective in demonstrating *that* Simpson's paradox can happen but limited in offering insight into *how* it can happen. Exploring additional representations can provide insight into the situation that will help resolve this tension. We will keep the underlying numbers the same, however, to keep the primary focus on the representation itself.

The Circle Graph Representation

One alternative representation is the circle graph (Paik 1985) as shown in figure 11.1. Each circle acts as a sort of scatterplot cluster of points representing one of the four gender-department combinations with the y-coordinate of each center at the corresponding hiring rate (.25, .375, .625, or .75). The "weighting" of each hiring rate is reflected in the area of each circle being proportional to the size (20, 80, 80, or 20, respectively) of the applicant pool for that particular gender-department combination. The top right circle and lower left circle are small, each representing a group of 20 people—physical sciences women or social sciences men, respectively. The larger circles represent groups of 80 people—physical sciences men and social sciences women. Since 80 = 4(20), one of the large circles should have quadruple the area (and double the radius) of a smaller circle in figure 11.1. (There is no significance to the absolute length of any particular radius nor to whether the endpoints of the line segment representing the "overall" situation lie inside the circles.)

The highest line segment in the interior of figure 11.1 connects the male and female hiring rates for physical sciences, and the lowest line segment does this for social sciences. Because females are (arbitrarily) placed to the right of males on the horizontal axis, the positive slopes of these line segments reflect the fact that females have a higher hiring rate than males within either department. The middle line segment connects the male and female hiring rates for the overall university, and we see that now (from its negative slope) it is males who have the higher hiring rate. The descriptions *highest, lowest,* and *middle* may not be sufficient to distinguish the three line segments for other data sets, such as if the departmental segments crossed each other (in such a situation, each department would favor a different gender).

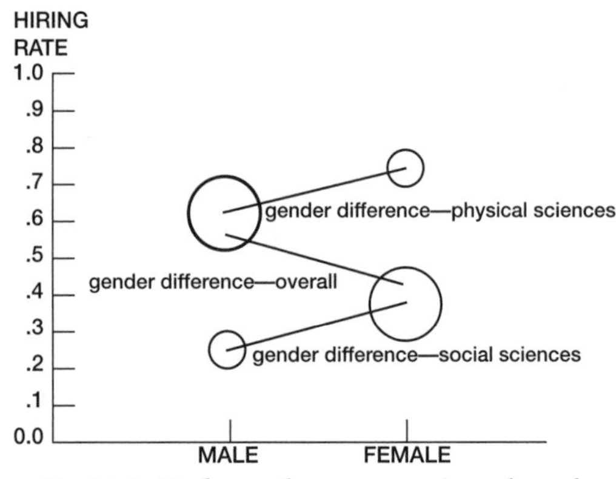

Fig. 11.1. Circle graph representation adapted from Paik (1985)

Although so far we have given an interpretation only to the sign of the slope, we can also interpret its numerical value as the "female minus male" difference in hiring rates if we choose to code the male and female markers on the x-axis as 0 and 1, respectively. With such coding, each department would have a segment with a slope of .125 while the (middle) segment representing all 200 applicants would have a slope of –.1. Coding a qualitative variable in such a "quantitative" manner is mathematically meaningful only when the qualitative variable is "dichotomous," which means having only two possible values (i.e., male or female). In addition, by similarly coding the dichotomous variable of hiring status (1 = "hired"; 0 = "not hired"), students who know how to calculate a correlation coefficient can verify that the correlation between gender and hiring status within either department is approximately .105 > 0, but for both departments combined is –.1 < 0, which is a connection to statistics showing a further representation of reversal. (The sign of a correlation coefficient is always the same as the sign of the slope of the line of best fit for that same scatterplot.)

When we look at the four circles, most of the "weight" (160 of the 200 applicants) is in the two large circles, whose positions determine a "negative sloping" orientation. The placement of the two smaller circles has a slight effect on this orientation, pulling the middle (overall) line segment somewhat counterclockwise (from the segment that would be determined by the centers of the two large circles alone), somewhat toward the orientation of the top and bottom lines, but not enough to attain a nonnegative slope. Students might try making circle graph sketches that vary the sizes or positions (i.e., heights) of the circles to suggest how all permutations (with replacement) of "positive slope," "negative slope," or "zero slope" could be possible for these three line segments for a new data set.

For example, if we change table 11.1 so that 60 of the 80 male physical sciences applicants and 20 of the 80 female social science applicants were hired, each department would produce a segment with zero slope (meaning males and females were hired at exactly equal rates within each department), but a disparity would still appear in the aggregate. This demonstrates that different "cell sizes" alone can cause the effect.

The Platform Scale Representation

Perhaps the most concrete representation besides the numerical table is the platform scale (see fig. 11.2), as described by Falk and Bar-Hillel (1980, p. 107):

> Suppose a set of uniform blocks arranged in stacks of varying heights is located on a weightless platform, which is balanced on a pivot located at the center of gravity.... One can ... shift the entire construction to the *right,* while simultaneously moving individual blocks to other stacks on their *left.* If done

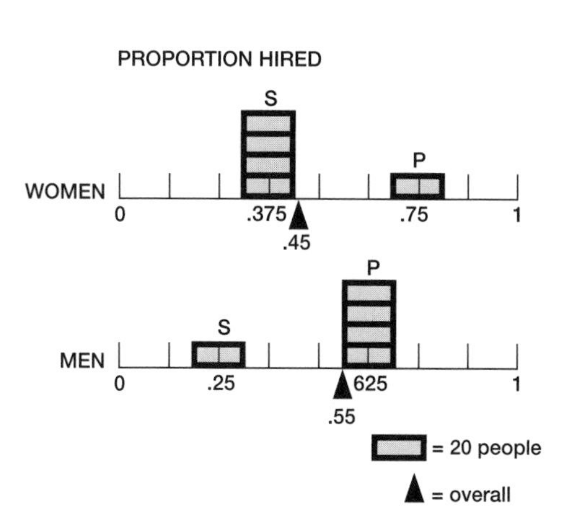

Fig. 11.2. Platform scale representation adapted from Falk and Bar-Hillel (1980)

appropriately, the net result could then be a new center of gravity which is to the left of the old one.

This helps students identify the weights and, we hope, recognize that weighted averages depend on the weights as well as on the values being averaged. The position of each stack represents an average, and the weight of the stack is the weight for that average (in computing the overall weighted average of averages). Students can certainly see with this representation that, for example, the weighted average and unweighted average of two stacks will be the same (i.e., have the same balance point) only if the sizes of the stacks are equal. In other words, if x and y are the stack position values and m and n are the stack weights, then setting $(nx + my)/(n + m)$ equal to $(x + y)/2$ yields $x(n - m) = y(n - m)$, which forces m and n to equal each other for $x \neq y$.

This representation builds naturally on intuition already provided by various textbooks, such as Billstein, Libeskind, and Lott (1993): "We can think of the mean as a balance point, where the total distance on one side of the mean (fulcrum) is the same as the total distance on the other side" (p. 459). Freedman et al. (1991) illustrate "histograms made out of wooden blocks attached to a stiff, weightless board. The histograms balance when supported at the average" (p. 59).

This representation could be readily extended to more than two stacks (departments, in this instance) and can be built with physical materials readily available in a typical classroom (e.g., using a ruler or meterstick for the platform). The platform scale representation may be limited, however, to numerical examples in which the weighting numbers (20, 80, 80, and 20) have a convenient greatest common divisor. Nevertheless, the intuition it provides would, it is hoped, give students intuition that could transfer to situations that cannot be as neatly modeled in this particular representation.

The Trapezoidal Representation

Tan (1986) provides a trapezoidal representation of Simpson's paradox (see fig. 11.3) that is built only on the observation (Hoehn 1984) that "[t]he length of any line segment which is parallel to the two bases and has its endpoints on the nonparallel sides of a trapezoid is the weighted mean of the lengths of the two bases" (p. 135). Specifically,

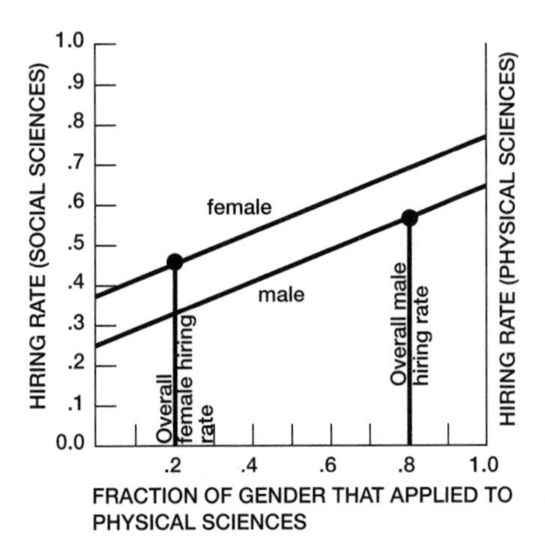

Fig. 11. 3. Trapezoidal representation adapted from Tan (1986)

each base is weighted by the proportion of the trapezoid's height traveled toward that base to reach the weighted mean segment. For example, the "female" trapezoid is determined by the points (0, 0), (0, .375), (1.0, 0), and (1.0, .75) and has bases of length .375 and .75, which have corresponding weights .8 and .2, respectively. We may conjecture that when the weights are each .5, the weighted average segment and the unweighted average segment coincide (in what geometry students would call the median of the trapezoid). This relationship can be verified algebraically by setting the usual formula for the area of the overall trapezoid equal to the sum of the areas of the two smaller trapezoids formed by the new segment, and then recognizing that we now have exactly the same equation we encountered when discussing the weighted and unweighted averages in the context of the platform scale representation.

Applying this to our university employment example, we find that each gender would have a trapezoid in which the two vertical bases represent that gender's hiring rates in the two departments. (This orientation is sideways from the more common depiction of a trapezoid, and the reason for this choice will soon be clear.) The trapezoids have one leg in common—the segment of the horizontal axis, which allows tracking the department "weights." The fact that females are hired at a higher rate than males in each department is clear by noting the vertical heights of the endpoints of the top male and female trapezoid legs (legs that happen to be parallel to each other but that do not have to be in general). However, because the genders had different proportions of applicants applying to the physical sciences department, it turned out that the overall male hiring rate (the large dot formed by extending the male departmental application proportion out to the top leg of the "male trapezoid") was higher (which is easier to see with the "sideways" orientation) than the large dot representing the overall female hiring rate.

Furthermore, it is straightforward to see with this representation when this reversal does and does not happen. For example, if the hiring rates for the four gender-department combinations were unchanged but the proportion of females who applied to the physical sciences became .8 (matching the proportion for males), then the dot representing the overall female hiring rate would slide along its line and clearly be higher than the dot representing the overall male hiring rate. Actually, we can solve the linear inequality $.75x + (.375)(1 - x) > .55$ to see that greater than 7/15 would suffice. This approach could also make it clear that when each gender has a 50-50 split of applications between departments, a reversal cannot occur. Related applications of this representation are made by Witmer (1992).

On the positive side, this model can be constructed physically (with a clip or bead sliding along a string strung between two vertical poles) or computationally (in interactive geometry software). These models would not be as limited to the data being "nice" as was required by the platform scale model. On the negative side, some students may be initially overwhelmed with the number of features of the graph, especially its having three axes.

The Unit Square Representation

Another type of geometric representation is adapted from the unit square model of Bea and Scholz (1994), who originally used it to represent conditional probabilities. Comparing shaded proportions of side-by-side rectangles (of equal length within each square), figure 11.4 shows at a glance that in each individual department the gender with the greater fraction hired is women and yet overall it is men. To understand better the representation, let's explain how the physical science unit square was constructed. Beginning with a square, draw a vertical line segment that partitions the square into rectan-

Proportions of Physical Sciences Unit Square by gender and hiring status

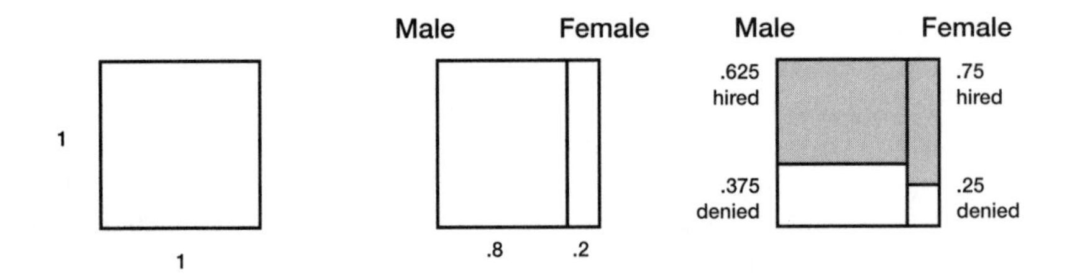

Proportions of Social Sciences Unit Square by gender and hiring status

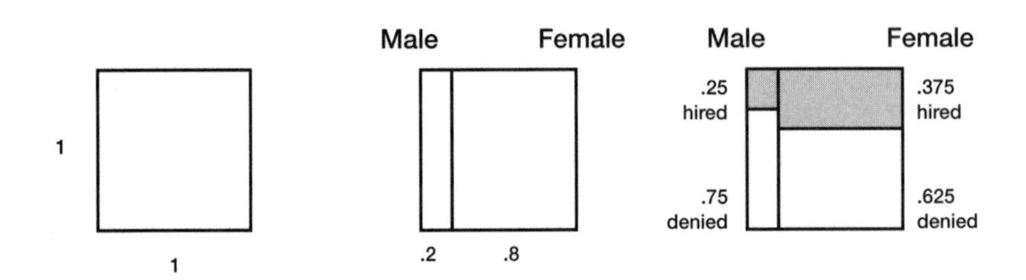

Proportions of Overall Unit Square by gender and hiring status

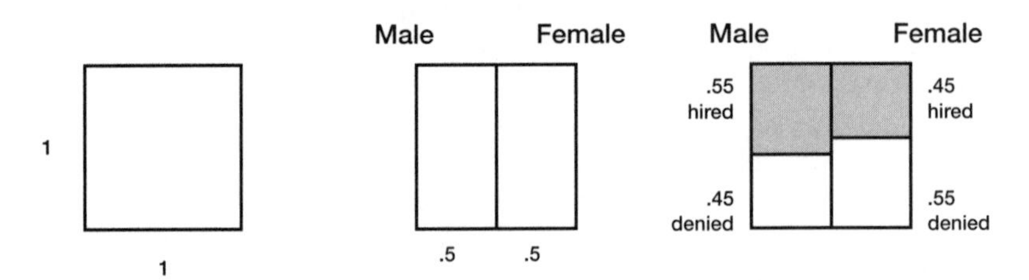

Fig. 11.4. Unit square representation of table 11.1 data adapted from Bea and Scholz (1994)

gles whose areas (and shorter sides) are proportional to the respective numbers of men and women (i.e., the vertical line would be 80/(20 + 80) of the distance from the left side to the right side). Now shade the fraction of each rectangle that represents the fraction of the corresponding gender that was hired (i.e., we shade 5/8 of the men's rectangle and 6/8 of the women's). Because the rectangles have the same height and because shading is done in the same direction, we see at a glance that the fraction of women hired was greater because its shaded region extends further down. Students who focus on the absolute amount of shaded area of each gender rather than the proportion of each rectangle that is shaded are simply noticing that within physical sciences, males had a greater *number* hired than females, but the *rate* at which men were hired is still less than the rate women were hired. This potential confusion (which would not occur in the other two unit square diagrams) offers a good opportunity to distinguish between amounts and rates.

Other Representations

There is also a probability representation (e.g., Movshovitz-Hadar and Webb 1998; Mitchem 1989) of this paradox that can be physically represented (and empirically simulated) in the classroom with four boxes and two colors of balls. Basically, the distribution of objects in the boxes is chosen so that the probabilities of drawing (with replacement) a certain color from box A or from box C are less than the probabilities of drawing that color from box B or from box D, respectively, but the inequality direction is reversed when the probability for the combined contents of boxes A and C is compared to the probability for the combined contents of B and D. Using the data from table 11.1, this would mean box A (labeled "social science males") would have 5 red chips and 15 green chips, box B ("social science females") would have 30 red chips and 50 green chips, box C ("physical science males") with 50 red chips and 30 green chips, and box D ("physical science females") having 15 red chips and 5 green chips. The phenomenon of combining data becomes very literal with this representation. Mitchem (1989) uses smaller numbers and a picture to shed further light on the dynamic.

Lord (1990) offers a representation involving determinants. Ignoring row or column totals in table 11.1, we can break the table into three 2 × 2 matrices, each of the following form:

$$\begin{bmatrix} \text{males hired} & \text{females hired} \\ \text{males denied} & \text{females denied} \end{bmatrix}$$

Recall that the determinant of the 2 × 2 matrix

$$\begin{bmatrix} a & b \\ c & d \end{bmatrix}$$

is $ad - bc$. Students may find it interesting to verify and investigate the fact that the determinant $(55)(55) - (45)(45) = 1000$ of the overall university matrix has a different sign than the determinant (–200) of either of the matrices representing an individual department. The reversal of determinant sign is not a coincidence. Students can verify that the male hiring rate being higher than the female hiring rate can be expressed as the inequality

$$\frac{a}{a+c} > \frac{b}{b+d}.$$

Algebraic transformations produce an equivalent inequality of $ad - bc > 0$, which corresponds precisely to a positive determinant for a 2×2 matrix! These manipulations also show that a determinant $ad - bc$ equal to 0 corresponds to the female hiring rate and male hiring rate being equal. Intuitively, then, the sign of the determinant tells us which gender has it better, and the larger the absolute value of the determinant, the greater the evidence of a statistical relationship or interaction between gender and employment status.

A more advanced representation uses vector geometry of the plane. Lord (1990, p. 55) demonstrates that "the following at-first-sight-plausible statement about complex numbers is, in fact, false: If $\arg(z_2) > \arg(z_1) > 0$ and $\arg(z_2') > \arg(z_1') > 0$, then $\arg(z_2 + z_2') > \arg(z_1 + z_1')$." This translates into a corresponding statement about slopes of vectors. Trying to make a diagram (such as fig. 11.5) that shows a specific counterexample is an interesting

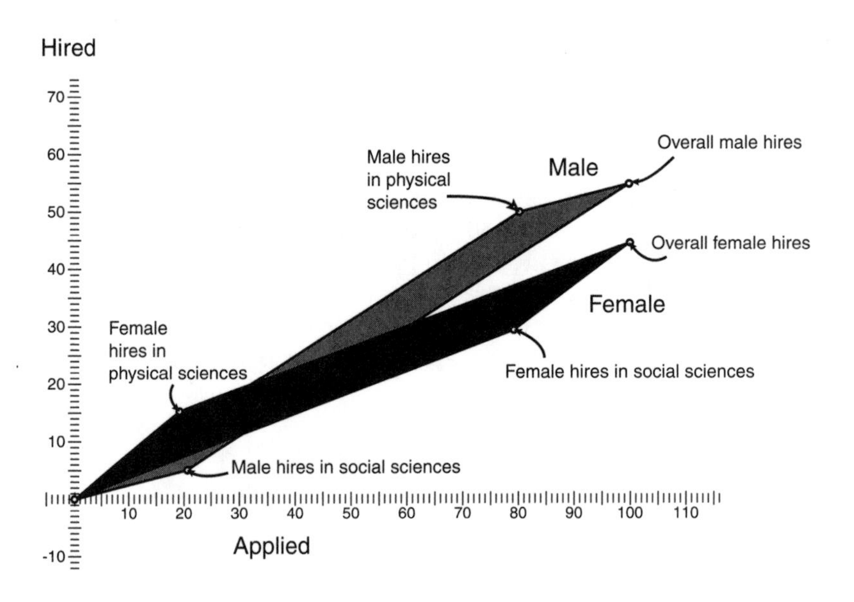

Fig. 11.5. A complex numbers representation of table 11.1

exercise. To facilitate the interpretation of our figure 11.5, let us agree that the ordered pair (a, b) now represents (number applied, number hired), so that slope corresponds to hiring rate. The physical science male vector goes from the origin to (80, 50), whereas the physical science female vector goes from the origin to (20, 15). The slope of the line segment or the angle relative to the horizontal axis made by the female vector is larger, thus showing that in the physical sciences, females have a better hiring rate. Similarly, the social science female vector has a slope larger than the corresponding male vector. But the overall male vector has a larger slope than the overall female vector.

An Exploration with Preservice Teachers

In the fall 1998 semester at a midsized, state-supported university, a class period of the secondary school mathematics methods course was devoted to exploring how multiple representations of a particular phenomenon served as tools for preservice secondary school teachers' thinking. Connections between mathematics content and pedagogy (e.g., using multiple representations) were one of several ways the author reformed this course as part of the National Science Foundation–funded Rocky Mountain Teacher Education Collaborative (Lesser 1999). All preservice teachers enrolled in the class participated, but the size of the class (seven) and time available for the exploration yielded anecdotal observations rather than definitive inferences.

Although all the students had had upper division coursework in probability and statistics, three of the seven initially answered that it was not possible "women could be hired at a higher rate than men within each of the two divisions, but still be hired at a lower rate than men for the university as a whole." To make sure that all students realized that, yes, it was possible, they were then shown table 11.1 and asked to verify by themselves the hiring rates for males and females within each department as well as for the overall university.

During the period, they had time to be exposed to seven of the representations previously mentioned (each with a written explanation of "how to read it" somewhat less detailed than is given in this article). Their comments made it clear that they had never seen most of the representations before in a textbook or class used to analyze any phenomenon, let alone Simpson's paradox in particular. There was a strong tendency among the students to say that although they might try the unit square or platform scale representations, they would most likely rely on the numerical representation in discussing the paradox with any future students. There are many possible explanations for students' reliance on the tabular (numerical) representation, including (1) it was the first one they saw, (2) it is the one most familiar to them from their mathematics classes, (3) it is the most concrete, (4) it is the one they could feel most confident teaching with, (5) it makes the most clear demonstration *that* the paradox occurs, and (6) they lacked the time to absorb others fully. The author believed the last explanation was certainly an issue. In fact, the next time he taught a secondary school methods course, a class period happened to be 2.5 times as long, and the students expressed a consensus that some of the physical and geometric representations were clearly superior to the table.

Biff, a preservice secondary school teacher in the author's methods course, addressed the tradeoffs of using multiple representations in general. On the one hand, he states: "[Multiple representations] can reach different kids with different approaches, and reinforce the learning—each representation is in fact teaching something new." The last part of this statement seems to suggest the recognition that a new representation is not just passively delivering the same piece of content but giving a new angle that may itself contain content (or even a transferable tool). Piez and Voxman (1997) believe that activities using multiple representations lead to more thorough understanding "[b]ecause each representation emphasizes and suppresses various aspects of a concept" (p. 164) and expect that "students will gain the flexibility necessary to work with a wide range of problems using an appropriate representation. In our work with students with weak mathematical skills, we have seen definite improvement" (p. 165).

However, Biff also expressed a pitfall: "Very time-consuming in the end. More sophisticated solutions will leave the slower in the dust." Going through all possible representations for each piece of content certainly would make it hard to stay on pace with a packed curriculum, but knowing that there may be far more than three representations possible makes it more likely to access the most useful one for the situation.

ISSUES AND IDEAS FOR THE CLASSROOM

The specific numbers chosen could be tailored to the audience. For example, there is more subtlety and "realistic appearance" in the data in table 11.2 (as opposed to table 11.1, whose entries are all multiples of 5 and included "swapped numbers").

TABLE 11.2
Slightly Revised Set of Hiring Data (by Gender and Department)

	Social Science		Physical Science	
	Male	Female	Male	Female
Hired	4	24	48	14
Denied	16	56	32	6

Constructing a data set with smaller numbers may make Simpson's paradox accessible even to students in the upper elementary grades and could even lend itself to being kinesthetically modeled by the typical number of students in a classroom by having students stand in marked-off regions of two 2 × 2 tables on the floor and then physically combine into a single 2 × 2 table. For example, a 19-student class could arrange themselves into two 2 × 2 tables as follows (1/3 < 3/8, 3/5 < 2/3, but 4/8 > 5/11):

1	3		3	2
2	5		2	1

If we admit the possibility of cells being 0, then Simpson's paradox can actually be physically modeled with as few as 9 students (0/1 < 1/4, 2/3 < 1/1, but 2/4 > 2/5):

0	1		2	1
1	3		1	0

The numerical representation is certainly the easiest way to introduce the phenomenon and can be presented in a very accessible manner through a structured sequence of questions (e.g., Smith 1996, p. 188) or as a story problem (e.g., Movshovitz-Hadar and Webb 1998, p. 113). In general, the way a particular representation is introduced may affect a student's ability to use or apply flexibly that representation. For example, in a project by McFarlane et al. (1995, pp. 476–77),

> children in the experimental classes were introduced to line graphs not as a Cartesian plot, where the ability to correctly identify positions on a grid was the objective; rather, they were introduced to graphs as a representation of the relationship between two variables.... Their ability to read and interpret temperature/time graphs was greatly enhanced as a result and it is particularly significant that their ability to sketch temperature time curves to predict the behavior of a novel system also improved.

Robust examples such as Simpson's paradox that have the potential to expand the repertoire of representations available may be especially valuable in the early part of the year to get students (and teacher!) primed to look for multiple ways of representing all future phenomena encountered, and to get them shaken out of habits they may have to fixate on nothing but one feature of some familiar graphical representation (see Berenson, Friel, and Bright 1993). More research in this area would be useful.

Some representations are certainly more suited toward technology representations than others. The numbers in the tables (as well as commands for the functions involved in any of the nonpictorial representations) could certainly be entered into a spreadsheet, for example, and the effect of changing various numbers instantly apparent. Shaughnessy, Garfield, and Greer (1996) list having a choice of dynamic representations for interpreting and displaying a data set to be an important attribute of a technological environment to facilitate the learning of data handling.

Simpson's paradox is rarely experienced by students in any of their courses, and if it were, it would likely be in numerical form only. Even a popular introductory college statistics textbook (Moore and McCabe 1993) that is quite "Simpson's paradox–oriented" (by involving the paradox in the only three-way table example in the text as well as in every three-way table exercise following that section) does not offer a nonnumerical representation of it. In a case study of a preservice secondary school teacher in a course integrating content and pedagogy, Wilson (1994) found that being able to translate between multiple representations was deeply related to conceptual understanding, a finding that supports Heid (1988) and NCTM (2000). Therefore, a deep understanding of how Simpson's paradox can occur seems difficult without the aid of representations beyond only a numerical one.

Perhaps exposing students and even teachers to rich representations of reversal representations will create a "reversal" of some of their attitudes, such as the common perception that representations are limited to context-free discussions of functions and are limited to a "rule of three" that they have already seen before! And maybe it will also reverse a perception of representations as some checklist of unrelated items to go through rather than as a dynamic source of new insights, connections, and tools for thinking whose roles should even further expand throughout the new century.

EXTENSIONS

As a follow-up assessment, students can be given the following data (table 11.3) and asked for a quick "gut" answer to the question "Is it possible that overall mean female salary is less than overall mean male salary even though mean female salaries are higher within each category?" This example may be a more subtle manifestation of the paradox in that the largest cell sizes for men and women are in the same category (i.e., support staff) this time, unlike table 11.1 (where 80 and 20 noticeably "swap roles" between departments). A "weighted average" computation shows the overall male and female salaries here are approximately \$41 000 and \$37 000, respectively.

TABLE 11.3
Annual Salary Data

	Men	Women
Support staff employees	70 males (their mean salary is \$20 000)	90 females (their mean salary is \$30 000)
Executive-level employees	30 males (their mean salary is \$90 000)	10 females (their mean salary is \$100 000)

Although Simpson's paradox may be new to students, they should be reminded that they have certainly seen inequality reversals earlier in their mathematical career, and they might now look for a representation that yields insight into those examples beyond "rules to memorize." For example, $2 < 4$ and yet $1/2 > 1/4$. An elementary school class might represent this with fair divisions of pizza, whereas a high school algebra class might consider the decreasing property of (either half of) the graph of $f(x) = 1/x$. Another "simple" example of reversal that students might look for a way to illuminate is why $-2 > -4$, since $2 < 4$.

For a final challenge, classes may look for a representation that indicates if it is possible to have the "double Simpson's paradox" posed by Friedlander and Wagon (1993, p. 268):

> It is possible for there to be two batters, Veteran and Youngster, and two pitchers, Righty and Lefty, such that Veteran's batting average against Righty is better than Youngster's average against Righty, and Veteran's batting average against Lefty is better than Youngster's average against Lefty, but yet Youngster's combined batting average against the two pitchers is better than Veteran's. ... [I]s it possible to have the situation just described [which would indeed be a feasible "single" Simpson's paradox] and, at the same time, have it be the case that Righty is a better pitcher than Lefty against either batter, but Lefty is a better pitcher than Righty against both batters combined?

If we adapted this to the employment context we have been working with throughout the article, the possibility of a "second" Simpson's paradox added to our scenario would correspond to asking, "Is it also possible that the social sciences department has a lower hiring rate than the physical sciences department for either gender and yet the physical sciences department has a lower hiring rate for both genders combined?"

REFERENCES

Bassett, Gilbert W. "Learning about Simpson's Paradox." *Stats* 12 (Fall 1994): 13–17.

Bea, Wolfgang, and Roland W. Scholz. "The Success of Graphic Models to Visualize Conditional Probabilities." Paper presented at International Conference on Teaching Statistics 4, Marrakesh, Morocco, July 1994.

Berenson, Sarah B., Susan Friel, and George Bright. "The Development of Elementary Teachers' Statistical Concepts in Relation to Graphical Representations." In *Proceedings of the 15th Annual Meeting of the International Group for the Psychology of Mathematics Education, North American Chapter,* Vol. 1, edited by Joanne Rossi Becker and Barbara J. Pence, pp. 285–91. Pacific Grove, Calif.: International Group for the Psychology of Mathematics Education, 1993. ERIC Document ED372917.

Bickel, Peter J., Eugene A. Hammel, and J. W. O'Connell. "Sex Bias in Graduate Admissions: Data from Berkeley." *Science* 187 (February 1975): 398–404.

Billstein, Rick, Shlomo Libeskind, and Johnny W. Lott. *A Problem-Solving Approach to Mathematics for Elementary School Teachers.* Reading, Mass.: Addison-Wesley, 1993.

Cohen, Joel E. "An Uncertainty Principle in Demography and the Unisex Issue." *American Statistician* 40 (February 1986): 32–39.

Falk, Ruma, and Maya Bar-Hillel. "Magic Possibilities of the Weighted Average." *Mathematics Magazine* 53 (March 1980): 106–7.

Freedman, David, Robert Pisani, Roger Purves, and Ani Adhikari. *Statistics.* 2nd ed. New York: W. W. Norton, 1991.

Friedlander, Richard, and Stan Wagon. "Double Simpson's Paradox." *Mathematics Magazine* 66 (October 1993): 268.

Heid, M. Kathleen. "Resequencing Skills and Concepts in Applied Calculus Using the Computer as a Tool." *Journal for Research in Mathematics Education* 19 (January 1988): 3–25.

Hemenway, David. "Why Your Classes Are Larger than 'Average.'" *Mathematics Magazine* 55 (May 1982): 162–64.

Hoehn, Larry. "A Geometrical Interpretation of the Weighted Mean." *College Mathematics Journal* 15 (March 1984): 135–39.

Lesser, Larry. "Countering Indifference Using Counterintuitive Examples." *Teaching Statistics* 20 (February 1998): 10–12.

Lesser, Lawrence Mark. "Investigating the Role of Standards-Based Education in a Pre-Service Secondary Math Methods Course." In *Promoting Excellence in Teacher Preparation: Undergraduate Reforms in Mathematics and Science,* edited by Myra L. Powers and Nancy K. Hartley, pp. 53–64. Fort Collins, Colo.: Colorado State University, 1999.

Lord, Nick. "From Vectors to Reversal Paradoxes." *Mathematical Gazette* 74 (March 1990): 55–58.

McFarlane, Angela, Yael Friedler, Paul Warwick, and Roland Chaplain. "Developing an Understanding of the Meaning of Line Graphs in Primary Science Investigations, Using Portable Computers and Data Logging Software." *Journal of Computers in Mathematics and Science Teaching* 14 (Winter 1995): 461–80.

Mitchem, John. "Paradoxes in Averages." *Mathematics Teacher* 82 (April 1989): 250–53.

Moore, David S., and George P. McCabe. *Introduction to the Practice of Statistics.* 2nd ed. New York: W. H. Freeman, 1993.

Movshovitz-Hadar, Nitsa, and Rina Hadass. "Preservice Education of Math Teachers Using Paradoxes." *Educational Studies in Mathematics* 21 (June 1990): 265–87.

Movshovitz-Hadar, Nitsa, and John Webb. *One Equals Zero and Other Mathematical Surprises.* Berkeley, Calif.: Key Curriculum Press, 1998.

National Council of Teachers of Mathematics (NCTM). *Professional Standards for Teaching Mathematics.* Reston, Va.: NCTM, 1991.

———. *Principles and Standards for School Mathematics.* Reston, Va.: NCTM, 2000.

Paik, Minja. "A Graphic Representation of a Three-Way Contingency Table: Simpson's Paradox and Correlation." *American Statistician* 39 (February 1985): 53–54.

Piez, Cynthia M., and Mary H. Voxman. "Multiple Representations: Using Different Perspectives to Form a Clearer Picture." *Mathematics Teacher* 90 (February 1997): 164–66.

Shaughnessy, J. Michael, Joan Garfield, and Brian Greer. "Data Handling." In *International Handbook of Mathematics Education,* Part I, edited by Alan J. Bishop et. al., pp. 205–38. Dordrecht, Netherlands: Kluwer, 1996.

Smith, Sanderson. *Agnesi to Zeno.* Berkeley, Calif.: Key Curriculum Press, 1996.

Tan, Arjun. "A Geometric Interpretation of Simpson's Paradox." *College Mathematics Journal* 17 (September 1986): 340–41.

Wagner, Clifford H. "Simpson's Paradox." *Journal of Undergraduate Mathematics and Its Applications* 3 (1982a): 185–98.

———. "Simpson's Paradox in Real Life." *American Statistician* 36 (February 1982b): 46–48.

———. "Simpson's Paradox." UMAP: Modules and Monographs in Undergraduate Mathematics and Its Applications Project, Unit 587. Newton, Mass.: Education Development Center, UMAP, 1983.

Wilensky, Uri. "Paradox, Programming and Learning Probability: A Case Study in a Connected Mathematics Framework." *Journal of Mathematical Behavior* 14 (1995): 253–80.

Wilson, Melvin. "One Preservice Secondary Teacher's Understanding of Function: The Impact of a Course Integrating Mathematical Content and Pedagogy." *Journal for Research in Mathematics Education* 25 (July 1994): 346–70.

Witmer, Jeffrey A. *Data Analysis: An Introduction.* Englewood Cliffs, N.J.: Prentice Hall, 1992.

12

Presenting and Representing
From Fractions to Rational Numbers

Susan J. Lamon

TEACHERS and researchers have known for many years that traditional instruction in fractions does not encourage meaningful performance from most students. It has been repeatedly documented in state and national assessments that students are learning little, if anything, about rational numbers. The current fraction curriculum consists of a specific set of procedures or algorithms for computational purposes that provide some basis for manipulating algebraic expressions but fails to help most children understand that the rationals, despite their bipartite character, are numbers in their own right.

This article discusses some of the persistent issues and problems surrounding fraction instruction, but it goes beyond documenting shortcomings to look at some promising alternatives to current practice. It examines some results from a longitudinal study of five classes of children, each of which began its study of fractions using an interpretation of the symbol *a*/*b* that differed from the standard "part out of the whole," and none of which was ever taught any computational algorithms. Finally, the article discusses issues and imperatives for current practice based on that study. As the title suggests, both *presentational* models (used by adults in instruction) and *representational* models (produced by students in learning) play significant roles in instruction and its outcomes.

To appreciate the broad base of meaning that is obscured when we are operating with fraction symbols, consider the following problems (solutions are at the end fo this article):

1. Does the shaded area in figure 12.1a show 1 (3/8 pie), 3 (1/8 pies), or 1 1/2 (1/4 pies)? Does it matter?

2. You have 16 candies. You divide them into 4 groups, select one group, and make it three times its size. What single operation would have accomplished the same result?

(a)

Fig. 12.1.

3. You have taken only one drink of juice, represented by the unshaded area in figure 12.1b. How much of your day's supply, consisting of two bottles of juice, do you have left?

4. If it takes 9 people 1 1/2 hours to do a job, how long will it take 6 people to do it?

5. Without using common denominators, name three fractions between 7/9 and 7/8.

6. Yesterday Alicia jogged 2 laps around the track in 5 minutes, and today she jogged 3 laps around the track in 8 minutes. On her faster day, assuming that she could maintain her pace, how long would it have taken her to do 5 laps?

7. Here are the dimensions of some photos: (*a*) 9 cm × 10 cm, (*b*) 10 cm × 12 cm, (*c*) 6 cm × 8 cm, (*d*) 5 cm × 6.5 cm, and (*e*) 8 cm × 9.6 cm. Which one of them might be an enlargement of which other one?

(b)

Fig. 12.1.

Although these problems are not mathematically difficult for someone who already has a robust understanding of rational numbers, they present many nuances that make them psychologically difficult from a teaching and learning perspective. Although we may have used fraction symbols to help us answer some of the questions, our underlying interpretation of the symbols was not always the part/whole interpretation: "*a* out of *b* equal parts of a unit." In fact, these problems address only a small sample of the rich meanings, relationships, and contexts that may be represented by a fraction symbol. Yet, schools have produced many generations of people whose only preparation for understanding the complex domain of rational numbers consists of a brief introduction to the part/whole interpretation, followed by years of lockstep practice in fraction computation. We are paying a high price for our long neglect of this part of the mathematics curriculum. Over the last ten years, 90 percent of the eighteen- and nineteen-year-old students in my mathematics classes have been unable to answer 50 percent of those questions. We can only conclude that they have not yet had enough experience to understand rational numbers. It is difficult to imagine that they will gain that experience by taking courses that assume a knowledge of rational numbers.

Ongoing research in university-level calculus classrooms (Pustejovsky 1999; Lamon forthcoming) suggests that students who begin university mathematics with only a part/whole interpretation of *a/b* may have missed their window of opportunity. Although poor algebra skills are most often blamed for the lack of success in calculus, we are finding that having little or no understanding of rational numbers accounts for most students' conceptual difficulties when trying to understand the derivative. Throughout a one-semester course, they show no understanding of derivatives and survive academically merely by learning patterns for calculating them. Students who begin with multiple interpretations of the symbols continue to develop meanings and connections that make their understanding more robust than this.

Students who are more comfortable when performing operations than they are when reasoning fail to make connections and are unable to transfer information to new situations, even when the transfer represents only a small leap. A dramatic example occurs in many trigonometry classes. When an angle is measured in radians and students are asked to convert the measurement to degrees, they usually need a rule and a calculator to get the job done. Consider the task of converting 5π/6 radians to degrees. Figure 12.2 shows two calculations for the conversion.

Fig. 12.2

Each of these is equivalent to finding 5/6 of 180, but calculation (b) is one that can be readily performed—that is, 5(1/6 of 180). Viewing the fraction 5/6 as an operator, that is, the composition of a division and a multiplication, makes the conversion an easy mental task (5 * 30). Although the mental operations and the calculator operations look similar, performing the operations on the calculator does not lead students to discover the mental conversion.

Similarly, it is surprising that students who divide a unit into fourths and designate 3 of them, obtaining a value of 3/4, cannot make the leap to recognize that when dividing 3 pizzas among 4 people, each will receive 3/4 of a pizza. The following example illustrates this failure. A nineteen-year-old first-semester calculus student was asked to consider the situation represented in figure 12.3. Three pizzas were to be divided among 7 girls, and 1 pizza was to be divided among 3 boys. The student was then asked, "Who gets more pizza, a girl or a boy?"

She replied: "I'm taking basically 7 divided by 3. That would be like"—she figured on her calculator—"2 point ... well, let's say 2 slices and maybe a third. And I knew that 3 divided by 1 would be"—again she figured on her calculator—"um, yeah, oh, it would be 3. But basically, the boys would get 1/3 of the pizza. So the girls would get less."

Fig. 12.3

This student, despite her part/whole training, showed no understanding of partitioning and fractions as quotients. She divided the number of people by the number of pizzas and interpreted her answers as the number of slices. Her intuitive knowledge contradicted her calculated answer—she knew that each boy would get 1/3 of a pizza—but she nevertheless went on to compare 2 1/3 slices to 3 slices. Her leap to equating "more pizza" with "more slices" is a strong indicator that she never understood part/whole comparisons in the first place. Her approach is similar to comparing the fractions 5/6 and 7/8 by observing that 7 > 5. It suggests a failure to distinguish the difference between the questions "How many?" and "How much?"

In short, there is ample evidence to question the efficacy of fraction instruction. Part/whole fractions as they are currently taught are not providing an adequate point of entry to the rational numbers, but mathematics education research has been slower in suggesting an alternative than in diagnosing students' deficiencies.

WHAT IS EFFECTIVE FRACTION INSTRUCTION?

The ultimate goal of instruction is to help students to understand fractions as numbers in their own right, and, as such, as objects that can be manipulated with arithmetic. Achieving that goal is not an easy task. It builds incrementally, over considerable time, as opposed to being an all-or-nothing occurrence. An early step toward the goal is acquiring the ability to understand relative comparisons and to abstract a single common notion across a wide range of representations within a single subconstruct. In the part/whole interpretation, this means that regardless of the representation used for the fraction and regardless of the size, shape, color, arrangement, orientation, and number of equivalent parts, the student can focus on the relative amount. For example, the number 1/4 describes the relative amount represented in the various shaded regions in figure 12.4. But this notion of equivalence is, by itself, very restricted. It depends only on the ability to divide objects into *n* equal parts. Mathematically and psychologically, the part/whole interpretation of fraction is not sufficient as a foundation for the system of rational numbers.

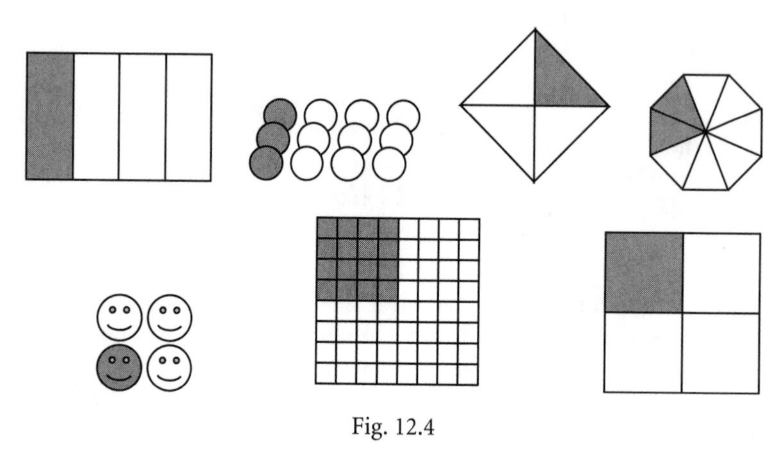

Fig. 12.4

The different interpretations of fractions that contribute to a robust understanding of a rational number have been summarized by Kieren (1976, 1980). He proposed that there are five different but interconnected subconstructs: part/whole comparisons, measures, operators, quotients, and ratios and rates. Each of these, in turn, has its own set of representations and operations, models that capture some—but not all—of the characteristics of the field of rational numbers. Even within one mode of representation that corresponds to a single interpretation—for example, a pictorial mode—there can be variation in the nuances of meaning, depending on whether discrete or continuous objects are used or whether unitary or solitary or composite objects are used, and so on. As the diagram in figure 12.5 suggests, basing instruction on a single interpretation and selectively introducing only some of its representations in instruction can leave the student with an inadequate foundation to support her or his understanding of the field of rational numbers.

Like many constructs in mathematics, the rational numbers can be understood only in a whole system of contexts, meanings, operations, and representations. This understanding entails the conceptual coordination of many mathematical ideas and results in flexible ways of thinking and operating, one of the most important being proportional reasoning. Vergnaud (1983, 1988) best captured the challenge and complexity of the mathematics entailed in the formidable task of knowing rational numbers

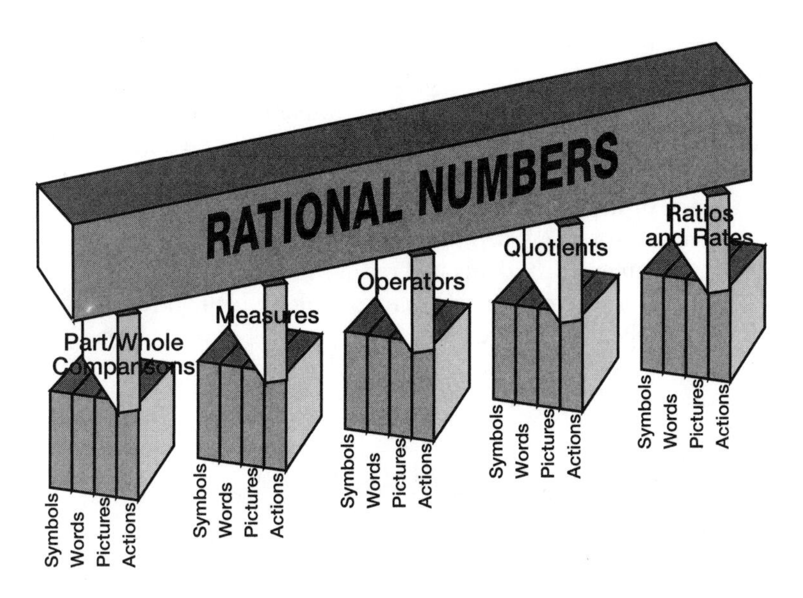

Fig. 12.5

when he described a conceptual field: a set of problems and situations for which closely connected concepts, procedures, and representations are necessary.

WHY DO WE TEACH PART/WHOLE FRACTIONS THE WAY WE DO?

Fractions and ratios were once the tools of clerks and bookkeepers. Ciphering was learned by meticulously copying the work of a master, and it had disciplinary as well as practical value. The computation of fractions assumed an important role in the elementary school mathematics curriculum with the expansion of business and commerce during the Industrial Revolution of the eighteenth and nineteenth centuries (National Council of Teachers of Mathematics [NCTM] 1970). By the early 1900s, psychology began to influence instruction, and it is easy to imagine what a great instructional innovation it was when part/whole relationships began to be used to introduce the vocabulary and symbolism of fractions. Still, students were afforded only a brief encounter with partitioning activities to build meaning because a time-efficient route to the formal symbolic computation best served the needs of society at that time.

Most current mathematical agendas favor some balancing of computational skills with a richer understanding that results in higher-order thinking and problem-solving skills. Nevertheless, instruction today is not very different from that of the nineteenth century. Some new curricula, developed within the last decade, afford students some access to meanings other than the part/whole interpretation, but there is evidence that long-term interaction with one or more of the subconstructs may be necessary. Studies done by researchers from the Rational Number Project (e.g., Behr et al. 1984) showed that even when students had received up to eighteen weeks of carefully planned, intensive instruction involving multiple interpretations of fraction symbols, they showed "a substantial lack of understanding" and could deal only with "questions that [did] not involve the application of fraction knowledge to a new situation" (p. 337).

WHAT DO WE HAVE TO BUILD ON?

As children begin to study fractions, they encounter many points of discontinuity with the whole-number system and its operations that require great cognitive leaps. (See Hiebert and Behr [1988] for a more complete discussion of these.) Nevertheless, children demonstrate some remarkable strengths on which instruction might build.

Just as children attend carefully to the actions and relationships expressed in addition and subtraction problems and then carefully model them without prior instruction (Carpenter and Moser 1983), their earliest fraction work demonstrates the same attention to problem structure. When asked to draw pictures for each of the following situations, third-grade children drew different representations for each:

1. Andy told his mom he would be home in three-quarters of an hour.

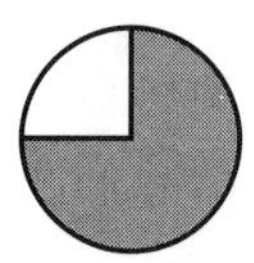

2. Mandy had 3 large cookies all the same size. One was chocolate chip, one was oatmeal, and one was a sugar cookie. She cut each cookie into 4 equal parts and then ate 1 piece of each.

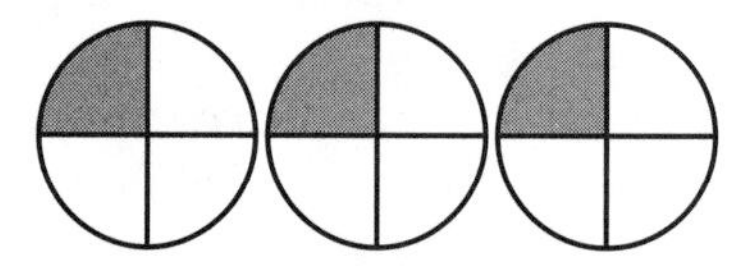

3. Three identical cupcakes are in a package. Marge cut the package into 4 equal parts without opening it. She gave you one of the parts.

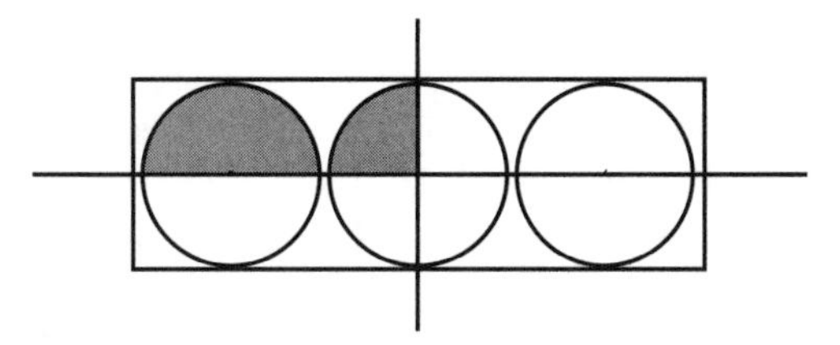

On an abstract level, the rational number 3/4 describes all these situations: 1(3/4 piece), 3 (1/4 pieces), and 1/4 of a 3-pack. Although these young students did not have the language to refer to the resulting amounts, their drawings and explanations made it clear that the situations were psychologically different. Students attended carefully to the meaning of each situation, and all situations were not equally effective in attaining the target concept. For example, they noted that Mandy's pieces in example 2 could not be mushed together because they were from different kinds of cookies, but when the problem was changed so that her cookies were all of the same type, they recognized that she had eaten an amount equivalent to 3/4 of a cookie.

These examples cover only a few of many different meanings that may be represented by the rational number 3/4. It may be a comparer, as in "There are only 3/4 as many men as women in the room." It may be a measurer, as in "3/4 bottle of juice." It may be a transformer, as when it shrinks something to 3/4 of its size. Helping children to recognize the multiple personalities of a rational number is a challenge, but their ability to detect nuances in the problems that they are given and to build meaning when they have no rules or algorithms are strengths that instruction might exploit.

Current instruction in fractions grossly underestimates what children can do without our help. They have a tremendous capacity to create ingenious solutions when they are challenged. For example, I recently visited a fifth-grade class to do some problem solving with the students. I showed them the pictures in figure 12.6 and told them that the first picture shows Jeb and Sarah Smart when they were younger, and the second shows them as they look now. I posed the following question: "Who grew faster between the first and second pictures, Jeb or Sarah?"

After discussing the situation for about five minutes, several groups came up with clever solutions, but one was memorable. Speaking for the

Fig. 12.6

group, a student explained: “Well, they both grew taller, but Jeb grew *taller.* Get it? He was shorter; then he grew taller. Well, actually, Sarah was shorter; then she grew taller, too. But he grew *TALLER.*” The student spoke loudly and with emphasis. When he realized that his words were not conveying what he meant, he showed some frustration. There was a long pause while he thought. Finally, he began again: “Look. I can show you.” He took a ruler and aligned it with the top of Sarah’s head *then* (in the first picture) and the top of her head *now* (in the second one). Then he did the same for Jeb’s head in the two pictures. “See,” he said. “The ruler is more slanty for Jeb. That proves he grew faster.”

Asking children challenging questions often reveals their fledgling status with more complex mathematical ideas, such as slope and rate of change, and equipped with this information, the teacher or researcher can plan the next step to push the student into new territory.

A Longitudinal Study

Several compelling questions suggested priorities for a research agenda that could eventually shape instruction and prepare students to understand rational numbers:

1. What understanding might develop if a subconstruct other than part/whole comparisons was used as the primary interpretation for instruction?
2. What sorts of instructional activities would support an understanding of each of the rational-number subconstructs not commonly included in the present curriculum?
3. How long does it take to develop a useful understanding of any single subconstruct of rational number, and how will we know when it is achieved? Because understanding is not all-or-none, at what point can we say that a student understands?
4. Are the subconstructs equally good alternatives for learning concepts and computation? Which will be robust enough to connect to other interpretations without direct instruction?
5. Are there any developmental processes or mechanisms operating as children build an understanding of rational numbers, or is there total incongruity between their knowledge of rational numbers and their knowledge of whole numbers?

A longitudinal study conducted by the author focused on children’s development of meanings and operations of rational numbers. For four years in two urban schools in different parts of the country, five classes in grades 3 through 6 experimented with building students’ understanding of rational numbers on a different initial interpretation of the symbol a/b. One class began its study of fractions by interpreting *a*/*b* as a part/whole comparison with unitizing. (Unitizing refers to the cognitive process of reconceptualizing a quantity according to some unit of measure that is convenient for thinking about, and operating on, the quantity. For example, the given unit in a fraction problem may be 24 colas, but a student may think about that quantity as 2(12-packs), 4(6-packs), 3(8-packs), or 1(24-pack).) The other classes interpreted *a*/*b* first as a measure, or as a quotient, an operator, or rates and ratios. A sixth class that received traditional fraction instruction served as a control group. This study presented a unique opportunity to compare unconventional approaches with traditional instruction and to document the sequencing and growth of ideas, the breadth and depth of the understanding that developed, and the connections that children were able to make with other subconstructs of the rational numbers.

Children were given the freedom and the encouragement to express their thinking in whatever manner they chose. The two teachers who taught the five classes facilitated learning through the kinds of activities they supported and the problems they posed. The groups were not taught any rules or operations.

The teachers sometimes used whole-class instruction, and there were occasional whole-group discussions of ideas arising from the activities. However, on most days, the mathematical activity consisted of group problem solving, reporting, and then individually writing and revising solutions for homework. The questions posed to the students were built on a content analysis of the future uses of rational numbers and the nuances in understanding that were needed to pursue higher mathematics through beginning calculus. A summary of the subconstructs, the kinds of activities that supported each of them, and some of the future connections to each is given in table 12.1. More detail about instructional activities can be found in Lamon (1999).

REPRESENTATIONS AND UNDERSTANDING

Because understanding is a moving target, it can always be better, stronger, more connected, deeper, or broader than it is currently. As we ventured into unknown territory at the start of this study, the teachers agreed with the author that they would judge whether or not students understood something according to the nature of their representations. If children individually construct knowledge, then there should be something unique about their representations and explanations; they should not look and sound exactly like those presented in instruction. For instruction, adults choose certain representations because they can see in them an embodiment of something they already understand—the representations are media for what the teacher already knows. It is easy for teachers to attribute magical powers to the

TABLE 12.1
Alternatives to Part/Whole Fraction Instruction

Rational Number Interpretations of 3/4	Meaning	Some Classroom Activities	Some Future Connections
Part/Whole Comparisons with Unitizing "3 parts out of 4 equal parts"	3/4 means three parts out of four equal parts of the unit, with equivalent fractions found by thinking of the parts in terms of larger or smaller chunks $\frac{3 \text{ (whole pies)}}{4 \text{ (whole pies)}} = \frac{12 \text{ (quarter pies)}}{16 \text{(quarter pies)}} = \frac{1\frac{1}{2} \text{ (pair of pies)}}{2 \text{ (pair of pies)}}$	• reasoning • generating equivalent names for quantities	rational number properties and operations
Operator "3/4 of something"	3/4 gives a rule that tells how to operate on a unit (or on the result of a previous operation); multiply by 3 and divide your result by 4 or divide by 4 and multiply the result by 3	• stretching and shrinking with machines and copiers • folding paper • using area models for multiplication and division	algebra of functions; multiplication of rational numbers
Ratios and Rates "3 parts to 4 parts" "per" quantities 3:4	3:4 means 3 parts of A are compared to 4 parts of B, where A and B are of like measure or 3 units of A per 4 units of B, where A and B are of unlike measure	• reasoning • using chips • making ratio tables • analyzing graphs • doing ratio arithmetic	rationals as equivalence classes of ordered pairs; operations and properties
Quotient "3 divided by 4"	3/4 is the amount each person receives when 4 people share a 3-unit of something	• partitioning sets of discrete and continuous objects	rational numbers as a quotient field
Measure "3 (1/4-units)"	3/4 means a disance of 3 (1/4-units) from 0 on the number line or 3 (1/4-units) of a given area	• successively partitioning number lines, areas, and volumes • reading meters and gauges	density of rational numbers; rational numbers as arbitrary divisions of a unit; vector operations

representations, thinking that the activity, the manipulative, the picture, or the words that have meaning for them will surely persuade students to adopt the teachers' adult perspective. Any good teacher knows that it just doesn't work that way! Students fit new ideas into their existing knowledge. When students give back on a test exactly what the teacher has presented in class, it is not clear at all that they have understood the material, and often the chances are good that they have *not* learned it. When a student truly understands something, in the sense of connecting or reconciling it with other information and experiences, the student may very well represent the material in some unique way that shows his or her comfort with the concepts and processes. The following examples make clear this distinction between genuine and surface learning.

In the rational-numbers-as-measure class, the students engaged in many activities in which they successively partitioned number lines, meters, and gauges until they could answer problems such as the following:

Locate 17/24 on the number line shown in figure 12.7.

Figure 12.8 shows how one student arrived at an answer. Students were taught to use arrow notation to keep track of the size of the subdivisions after each time they partitioned.

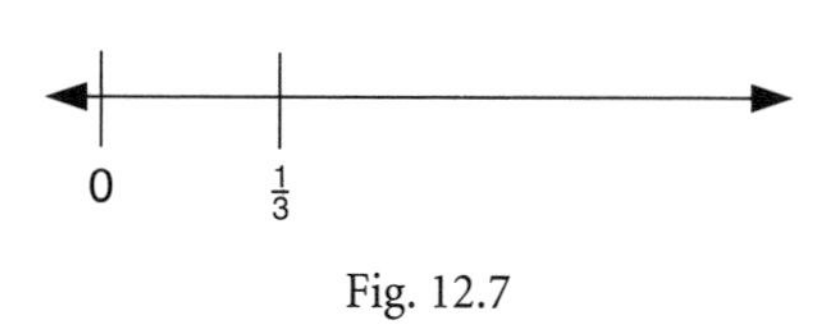

Fig. 12.7

Students' use of arrow notation decreased over time; they gradually stopped partitioning the number lines and meters and were able to visualize

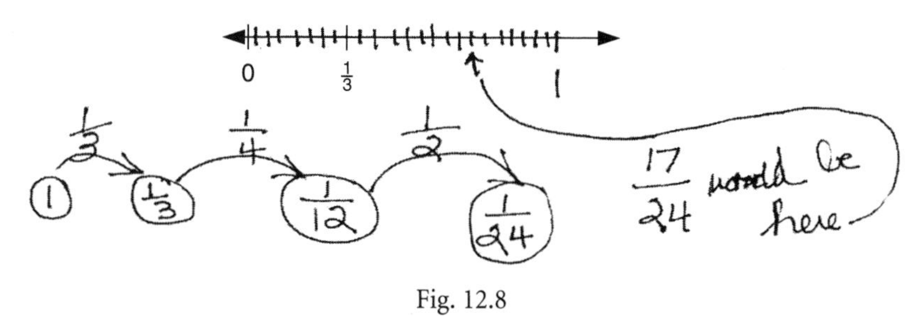

Fig. 12.8

and reason about the fractions without drawing the number lines. Although we felt that this process indicated progress in a positive direction, our criterion for student understanding was original representations rather than those that mimicked the ones presented in instruction.

After a long time—sometimes more than two and a half years—students began to show extraordinary comfort and adaptability in dealing with the questions posed. One of the most remarkable understandings these children developed was a sense of the density of the rational numbers. Figures 12.9 and 12.10 show

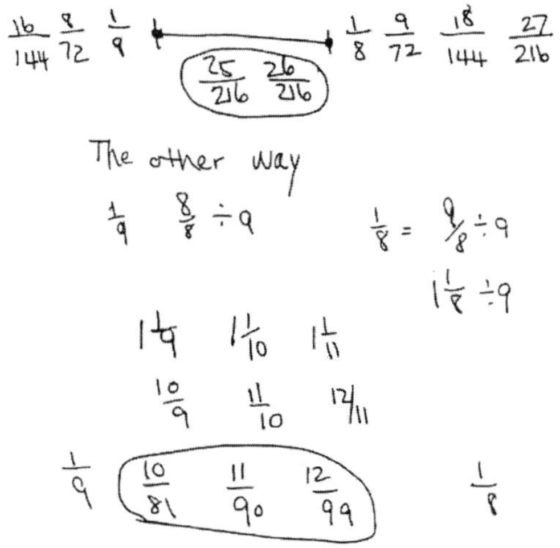

Fig. 12.9

two of the strategies the children produced to find fractions "between" two given fractions. Figure 12.9 shows the work of a student at the end of grade 5. The teacher asked him to find two fractions between 1/8 and 1/9. In explaining his work, the student said: "The teacher didn't say if she wanted the fraction equally spaced or not. If you don't care if all the fractions are equally spaced or not, I could find lots of other fractions between 1/9 and 1/8. If I pick something less than 1 1/8 and divide it into 9 equal parts, it will be between those two fractions. Like I could pick 1 1/9, 1 1/10, 1 1/11, 1 1/12, or as many as I want, ... divide 'em by 9 ... and they'll all be in there."

Figure 12.10 shows the work at midyear of a sixth grader who was asked to find two fractions between 7/12 and 7/11. We considered both this example and the previous one to be true representations. The students had gone well beyond the successive partitioning activities with which instruction began. Their understanding was individually constructed and expressed. Their knowledge of the relative sizes of rational numbers was useful in their problem solving, and many students had devised algorithms for producing any number of fractions that might be requested.

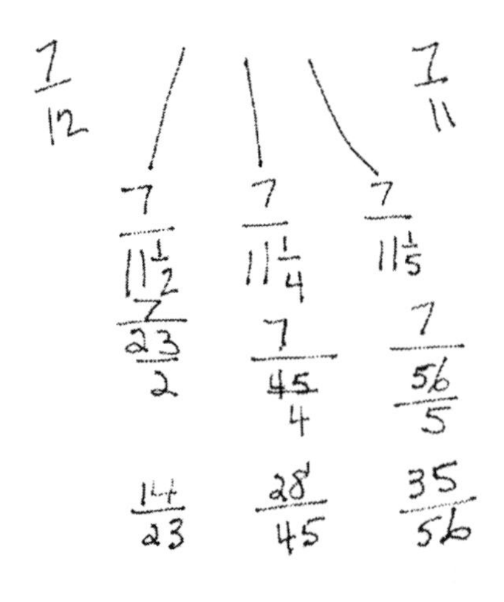

Fig. 12.10

Children in the class that started with the ratio-and-rates interpretation of rational numbers began by discussing what changed and what didn't change (variance and invariance) in situations involving two extraterrestrials who consumed given numbers of food pellets during given numbers of days. A sample situation is shown in figure 12.11.

As the numbers of food pellets varied, the children began to use poker chips to help model the problems and answer the question "Who ate more, Snake Woman or Slime Man?" They adopted the term *cloning* for making exact copies. For example, they called

a 3-clone of

because they look alike, they act alike, and you can use one in place of the other. With the notion of equivalence well in place by the end of third grade,

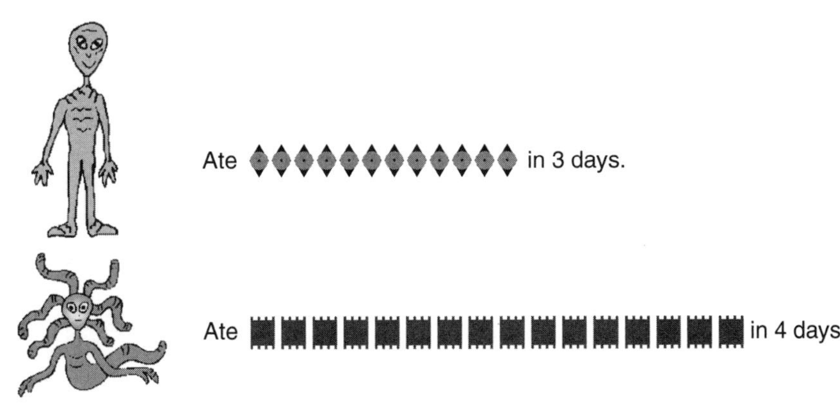

Fig. 12.11

these children developed sophisticated strategies for comparing rations and rates during the fourth and fifth grades. For instance, consider Nancy's response to the following problem:

> Slime Man eats 5 food pellets in 7 days. Snake Women eats 3 food pellets in 4 days. Who is getting more food?

Nancy produced the following representation and explained what to do:

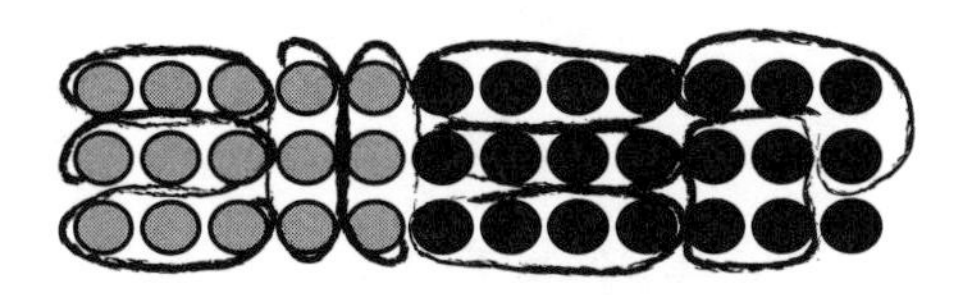

Nancy: Clone 5:7 and subtract clones of 3:4.

3(5:7) – 5(3:4)
15:21 – 15:20 = 0:1

No matter how many times you clone 5:7 and take 3:4 out of it, you are always going to have more days than you have food. So 3:4 must mean more food.

The evolution of the children's representations, including language and chip formations, neither of which was presented by the teacher, suggested that the operations they were performing resulted from their deep understanding of ratios.

SELECTED RESULTS

After four years, all five groups of students had developed a deeper understanding of rational number than the students in the control group had, as measured by the number of subconstructs that the students were using. The numbers of proportional reasoners in the five classes far exceeded the number in the control group, and even in computation, achievement was greater in all five groups than it was in the control group.

Using different interpretations of rational numbers, children acquired meanings and processes in different sequences, to different depths of understanding, and at different rates. Even children who were taught the same initial interpretation of fractions showed different learning profiles. The time-honored learning principle of transferability was robust. Children transferred their knowledge not only to unfamiliar circumstances but also to other interpretations that they had not been directly taught. By the end of sixth grade, more than 50 percent of the children had demonstrated the ability to apply their knowledge to at least two of the rational-number subconstructs. Table 12.2 shows the numbers of students from each class who were able to reason proportionally at the end of four years, the number from each class who could pass a computation test at the 80 percent level, and the number from each class who had demonstrated the ability to apply their knowledge to as many as four of the rational-number subconstructs.

Children in all five groups learned to reason with fractions. They never received rules or algorithms but expected to be given new challenges and to engage in problem solving. This approach was routine in their mathematics

TABLE 12.2
Achievement of the Four-Year Participants from Each Class on Three Criteria

	Numbers of Students Achieving Each Criterion						
			Number of Rational Number Interpretations				
Class	Proportional Reasoning	Computation[a]	1	2	3	4	5
Unitizing[b]	8	13	19	8	7	1	0
Measures[c]	8	12	17	11	9	3	0
Operators[d]	3	9	15	9	0	0	0
Quotients[e]	2	9	16	7	0	0	0
Ratio/Rate[f]	12	11	18	10	6	0	0
Control[g]	1	6	11	2	0	0	0

[a]80% accuracy or better.
[b]$n = 19$.
[c]$n = 17$.
[d]$n = 19$.
[e]$n = 18$.
[f]$n = 18$.
[g]$n = 20$.

classes. Although they had never engaged in formal computation, some fifth graders (approximately at midyear) were asked to explain how to solve the problem 1 ÷ 2/3. After more than two years of instruction, they were such flexible thinkers that they produced a wide variety of solutions. Figure 12.12 shows three students' representations of their thinking.

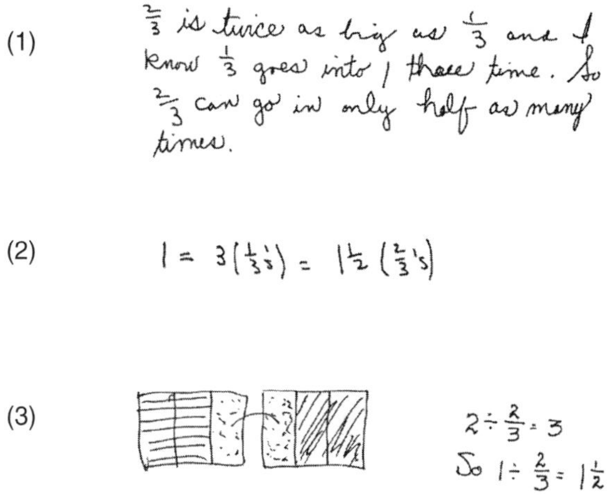

(1) $\frac{2}{3}$ is twice as big as $\frac{1}{3}$ and I know $\frac{1}{3}$ goes into 1 three time. So $\frac{2}{3}$ can go in only half as many times.

(2) $1 = 3(\frac{1}{3}\text{'s}) = 1\frac{1}{2}(\frac{2}{3}\text{'s})$

(3) $2 \div \frac{2}{3} = 3$
So $1 \div \frac{2}{3} = 1\frac{1}{2}$

Fig. 12.12

After four years, students' reasoning and problem solving in the five special classes were extraordinary. Their abilities called into question the conclusion from former, now classic, research studies that the type of numbers used affects the solution of fraction or ratio comparison tasks. For example, consider the following comparison task:

Are the fractions 3/5 and 7/11 equivalent?

Adults may use a common denominator or check cross products to see that $3/5 < 7/11$. When elementary school students in the beginning stages of fraction instruction are asked this question, most conclude that the fractions are not equivalent because there is no natural number by which 3 can be multiplied to get 7 and 5 can be multiplied to get 11. They will, in fact, arrive at the same conclusion for the same reason when asked to compare 4/6 and 6/9. Older children who have learned how to reduce or to find common denominators do a little better. However, middle school students generally perform very poorly on this type of comparison. (For example, see Karplus, Pulos, and Stage 1983). Either children do not know how to proceed, or they resort to incorrect additive strategies. Researchers have concluded that the lack of integral multiples between and within the given ratios greatly contributes to their difficulty. Middle school students generally find it easier to compare fractions or ratios such as 3/12 and 5/24, 6/8 and 12/15, or 3/9 and 6/15.

When our children were given all the tasks used in the study by Karplus, Pulos, and Stage, they made no more errors on comparison tasks without integral multiples than they made on tasks involving integral relationships. Figure 12.13 shows two responses to the question "Are 3/5 and 7/11 equivalent?"—a comparison that is supposedly one of the most difficult types.

(1)

$$\frac{3 \text{ pies}}{5 \text{ pies}} = \frac{21\ (\frac{1}{7} \text{ pies})}{35\ (\frac{1}{7} \text{ pies})}$$

$$\frac{7 \text{ pies}}{11 \text{ pies}} = \frac{21\ (\frac{1}{3} \text{ pies})}{33\ (\frac{1}{3} \text{ pies})}$$

The first person and the second person have the same number of pieces but the first one has smaller pieces so he has less pie.

(2)

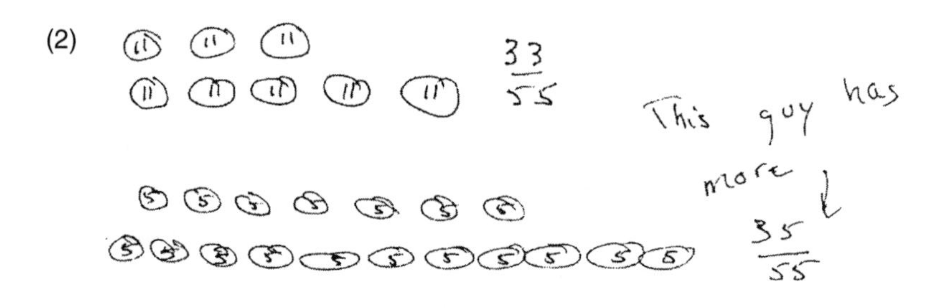

Fig. 12.13

Our results suggest that the kind of instruction children receive may have more impact on their ability to compare these fractions than the kind of numbers involved. The children in the study by Karplus, Pulos, and Stage were in sixth and eighth grades and presumably had had traditional, part/whole instruction. When rules and procedures are not learned with meaning, students forget them or do not always realize when to use them. However, when children are accustomed to thinking and reasoning without rules, which numbers they are given makes little difference.

IMPLICATIONS

It is likely that current part/whole fraction instruction could be improved merely by allowing children the time and opportunity to build understanding without directly presented rules and algorithms. In our study, it took a long time to build meaning, and during the first two to two and a half years, our students could not compete in fraction computation with children who had been using the traditional algorthms. However, in the long run, all five classes of students surpassed the rote learners, and their representations showed that they were performing meaningful operations. Of course, this delayed gratification has many implications for teacher accountability, state and national assessment, and many other aspects of schools and schooling.

Nevertheless, there is a case to be made for considering more dramatic changes in fraction instruction. It is apparent that when the goal of instruction is to provide as broad and as deep a foundation as possible for meanings and operations with rational numbers, not all the subconstructs of rational numbers are equally good starting points. Part/whole instruction as it is currently delivered was the least valuable road into the system of rational numbers. The part/whole interpretation with unitizing and the measure interpretation were particularly strong. Both rely heavily on the principles of measurement—the inverse relationship between the size of the unit with which one is measuring and the number of times one can measure it, out of a given quantity, and the successive partitioning of that unit of measure into finer and finer subunits until one can name the amount in a given quantity. The need for increasingly accurate measurement has been a driving mechanism in the history of mathematics and science, and it appears to be a strong binding force in children's mathematical development. The unitizing and measure interpretations and their corresponding representations show exceptional promise for helping children make the transition from whole numbers to rational numbers because these interpretations build on and extend principles of measurement with which the children have been familiar since early childhood.

It is not yet clear how the five interpretations can or should be integrated in instruction. It seems clear that a cursory look at each of them is unlikely to be of any real value in light of the long time it took students to grow comfortable with the single interpretation with which they began fraction instruction. It is also not clear whether computation should be developed in parallel to meaning. For the children in this study, meaning preceded and suggested appropriate operations. For example, the arithmetic that developed in the ratio-and-rates class said that (3:4) + (2:5) = (5:9) and that 3(3:5) = 9:15, whereas in other classes,

$$3\left(\frac{3}{5}\right) \neq \frac{9}{15} \text{ and } \frac{3}{4} + \frac{2}{5} \neq \frac{5}{9}.$$

It is not yet known to what extent introducing rules for operations might interfere with the development of meaning, or what might be gained by doing so.

Although there are still many researchable issues, we have a much better notion than we had twenty years ago of what it means to teach fractions for meaning. We also have a much better understanding of the role that presenting and representing fractions in instruction plays in enabling or disabling the development of rational-number understanding.

REFERENCES

Behr, Merlyn J., Ipke Wachsmuth, Thomas R. Post, and Richard Lesh. "Order and Equivalence of Rational Numbers: A Clinical Teaching Experiment." *Journal for Research in Mathematics Education* 15 (November 1984): 323–41.

Carpenter, Thomas P., and James M. Moser. "Acquisition of Addition and Subtraction Concepts." In *Acquisition of Mathematics Concepts and Processes,* edited by Richard Lesh and Marsha Landau, pp. 7–44. New York: Academic Press, 1983.

Hiebert, James, and Merlyn Behr, eds. *Number Concepts and Operations in the Middle Grades.* Reston, Va.: National Council of Teachers of Mathematics, 1988.

Karplus, Robert, Steven Pulos, and Elizabeth K. Stage. "Proportional Reasoning of Early Adolescents." In *Acquisition of Mathematics Concepts and Processes,* edited by Richard Lesh and Marsha Landau, pp. 45–90. New York: Academic Press, 1983.

Kieren, Thomas E. "On the Mathematical, Cognitive, and Instructional Foundations of Rational Numbers." In *Number and Measurement,* edited by Richard A. Lesh, pp. 101–44. Columbus, Ohio: ERIC-SMEAC, 1976.

———. "The Rational Number Construct: Its Elements and Mechanisms." In *Recent Research on Number Learning,* edited by Thomas E. Kieren, pp. 125–49. Columbus, Ohio: ERIC-SMEAC, 1980.

Lamon, Susan J. *Teaching Fractions and Ratios for Understanding: Essential Content Knowledge and Instructional Strategies for Teachers.* Mahwah, N.J.: Lawrence Erlbaum Associates, 1999.

———. "Understanding Derivatives: The Role of Rational Number Knowledge." Forthcoming.

National Council of Teachers of Mathematics (NCTM). *A History of Mathematics Education in the United States and Canada.* Thirty-second Yearbook of the NCTM. Washington, D.C.: NCTM, 1970.

Pustejovsky, S. F. "*Beginning Calculus Students' Understanding of the Derivative: Three Case Studies.*" Doctoral diss., Marquette University, 1999.

Vergnaud, Gerard. "Multiplicative Structures." In *Acquisition of Mathematics Concepts and Processes,* edited by Richard Lesh and Marsha Landau, pp. 127–74. New York: Academic Press, 1983.

———. "Multiplicative Structures." In *Number Concepts and Operations in the Middle Grades,* edited by James Hiebert and Merlyn Behr, pp. 141–61. Reston, Va.: National Council of Teachers of Mathematics, 1988.

Solutions to problems 1–7

1. Yes, it matters. The picture shows 3 (1/8 pies). Children cognitively differentiate the three quantities named here. At the abstract symbolic level, the quantities are indistinguishable, but in instruction, where contexts should be used to build an understanding of fractions, saying that these are the same contradicts children's intuitive understanding.

2. Dividing by 4 and then multiplying the result by 3 is the same as multiplying by 3/4.

3. The 1/5 bottle that you drank is 1/10 of your juice supply, so 9/10 remains.

4. If the number of people is 2/3 the original number, the time is 3/2 the original time because the number of people working and the time it takes to do a job are inversely related. It will take 6 people 2 1/4 hours to do the job.

5. A number of strategies are possible here. See some of the student strategies in this article.

6. Yesterday, Alicia jogged 2/5 of a lap per minute, and today she jogged 3/8 of a lap per minute. At yesterday's rate, it would take her 12 1/2 minutes to do 5 laps.

7. Ratios of corresponding sides show that (*b*) is an enlargement of (*e*).

13

Representations of Patterns and Functions
Tools for Learning

Wendy N. Coulombe

Sarah B. Berenson

In doing mathematics, we use various representations of problems and ideas to communicate our thinking to ourselves and to others. Indeed, it can be said that in many ways, representation *is* the language of mathematics. If we think of representation not only as an image (e.g., graph, table, diagram) but as a *process* of the illumination of ideas, we can see its usefulness in the learning of mathematics. Specifically, interpretation and translation of representations are tools that can extend students' algebraic thinking by helping students to construct their mental images of patterns and functions (Moschkovich, Schoenfeld, and Arcavi 1993).

Students of traditional algebra begin their formal study of functions by translating symbolic representations to tabular and then to graphical representations. A missing piece in this traditional approach is the students' interpretation of the algebraic symbols or other representations from a familiar context. Figure 13.1 illustrates two approaches to teaching equations with two variables. In most traditional approaches, a symbolic representation is given to the students, who then must translate this representation to a tabular and then to a graphical representation. In a problem-based approach (Janvier 1987), students are asked to interpret a commonplace scenario, such as that of saving money or dieting. The initial representation may draw from a variety of well-known representations. For example, a distance/time graph of two cars can be an initial representation. Students then interpret and translate the distance/time graph to write a description that compares the speeds of the two cars. A problem-based approach provides students opportunities for graphical construction and interpretation, data generation, pattern finding, and other interpretive processes (see Janvier [1987]).

Traditional Approach			Problem-Based Approach*		
Move From	**To**	**Translation Process**	**Move From**	**To**	**Interpretive Process**
Symbols ($y = 2x$)	Table	Compute values	Graph	Words	Graphical interpretation
Table	Graph	Plot points	Table	Words	Verbal pattern description
			Words	Graph	Qualitative graphical construction
			Graph	Table	Data generation
			Table	Symbols	Pattern finding

*Additional possibilities exist for problem-based lessons

Fig. 13.1. A comparison of two approaches used to improve students' understanding of functions (adapted from Janvier 1987)

Once given the opportunity to interpret familiar events mathematically, the representations take on a deeper meaning to students. The process of interpretation allows them to connect important mathematical ideas such as covariation and rate of change to the representations. Without interpretations, students' use of conventional mathematical representations may become an exercise in manipulating formulas to build tables and plot points on a graph.

The National Council of Teachers of Mathematics (NCTM) suggests that high school students should be able to model many different contexts using a variety of representations (NCTM 2000). This suggestion assumes that students can interpret different phenomena from the physical and social world and represent the relevant relationships. Inherent in the notion of interpretation is the idea that students can communicate their interpretations to others. According to NCTM (2000), communication is fundamental to all learning and is inseparable from representation. For this particular discussion, we will consider students' oral and written language both as a tool for communication and as a form of mathematical representation.

LEARNING FROM PIECEWISE-DEFINED FUNCTIONS

The learning of piecewise-defined functions is an interesting example of how the interpretation and translation of various representations can improve students' understanding. Piecewise-defined functions occur naturally and frequently within a wide range of physical and social environments in our world. In a piecewise-defined function, the domain is separated into two or more pieces, and a different functional relation is defined for each piece, as shown (symbolically) in figure 13.2.

The symbolic expression of piecewise-defined functions is troublesome even to students in precalculus and calculus courses. However, research suggests (Coulombe 1997) that first-year algebra students have intuitive notions about context-based piecewise-defined functions. The interpretation of graphical, tabular, and verbal representations of piecewise-defined functions can provide students with rich experiences to broaden their conceptualizations of patterns and functions.

$$f(x) = \begin{cases} 4, & x \le 6 \\ x - 2, & 6 < x < 9 \\ 7, & x \ge 9 \end{cases}$$

Fig. 13.2. An example of a symbolic expression of a piecewise-defined function

The Weight-Loss Problem

For example, first-year algebra students were asked to generate appropriate data values based on their interpretation of the qualitative graph in figure 13.3. In translating from a qualitative graph to a numerical (quantitative) table, the students interpreted the covariation of the independent and dependent variables and the various rates of change that were represented in the graph. The function involved was a linear piecewise-defined function, which appeared to be divided into three equal pieces. The first "third" could be seen to be constant, the second "third" was a negatively sloped line segment, and the final "third" was also constant. On the basis of the information provided in the table, one could assume that Kelly's weight remained stable at 133 during the first "third" of the time, and then after a loss of 3 pounds, remained stable at 130 during the last "third" of the time.

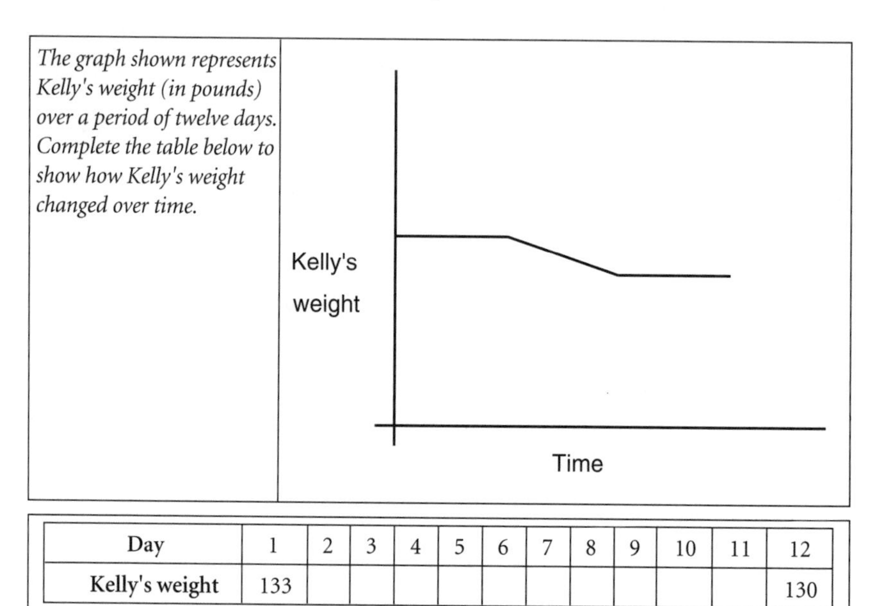

Day	1	2	3	4	5	6	7	8	9	10	11	12
Kelly's weight	133											130

Fig. 13.3. A data generation problem about weight loss

In the table shown in figure 13.4, Mac represented Kelly's weight remaining stable over a period of time at 133, then decreasing at a constant rate over two days, and then remaining stable at 130. In Anna's and Alex's tables, however, Kelly's weight does not decrease monotonically—they both show Kelly's weight remaining constant at 132 and 131 for more than one day. In addition, Alex underestimated the length of time that Kelly's weight remained constant at 130. Alternative responses such as Lynn's included table values that did not represent the constant portions of the graph, did not decrease at a constant rate, or were outside of the range of weights provided (130–133).

Day	1	2	3	4	5	6	7	8	9	10	11	12
Mac	133	133	133	133	133	132	131	130	130	130	130	130
Anna	133	133	133	133	132	132	131	131	130	130	130	130
Alex	133	133	133	132.5	132	132	131.5	131	131	130.5	130	130
Lynn	133	132	131.4	131.3	131.2	131.1	130.5	130.4	130.3	130.2	130.1	130

Fig. 13.4. Student responses to the weight loss problem

The Iced-Tea Problem

The problem in figure 13.5 gives students an opportunity to interpret, compare, and contrast three positive rates of change. Students can explain that at first, the tea was very sweet because Lily added four tablespoons of sugar for every two quarts of tea. Then the tea was less sweet because she

Lily needed to make iced tea for her soccer team in a 12-quart jug. She added black pekoe tea and sugar in the amounts shown in the table. Describe the sweetness of the iced tea at various points in her mixing.

Black pekoe tea (quarts)	Sugar (tablespoons)
0	0
2	4
4	8
6	10
8	12
10	15
12	18

Fig. 13.5. A verbal pattern description problem about mixing iced tea

added only two tablespoons for every two quarts of tea. Then she must have decided that three tablespoons for every two quarts was just right. For example, we might expect a student/teacher dialog similar to the following:

Student: The tea is more sweet in the beginning because she used four tablespoons in two quarts instead of two tablespoons in two quarts.

Teacher: So that means it's sweeter because ...

Student: Because four tablespoons in two quarts compared to two tablespoons in two quarts is sweeter.

Teacher: And why is it sweeter?

Student: Because it's different amounts of sugar and the same amount of tea. Tea is on a constant interval.

The Allowance Problem

The problem in figure 13.6 asks students to model three rates of savings growth over time. The graphs in figure 13.7 provide three responses to this problem that are typical of algebra students. Corey's graph illustrates the amount of money rising steadily, remaining constant, and then rising steadily again, although more quickly than it had in the beginning. Luke represents the amount of money rising first at a fairly constant rate and then later at a more rapid rate, but he omits the effect of the lack of allowance received in the spring. The points plotted in Elisabeth's graph depict the amount of money changing as a series of isolated events, even when there is no fluctuation in the amount of savings during the spring.

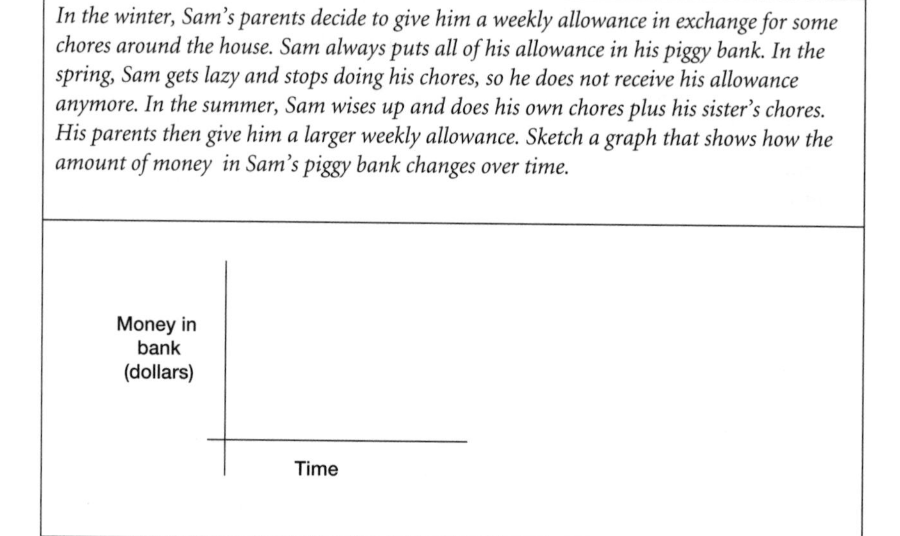

Fig. 13.6. A qualitative graphical construction problem about saving money

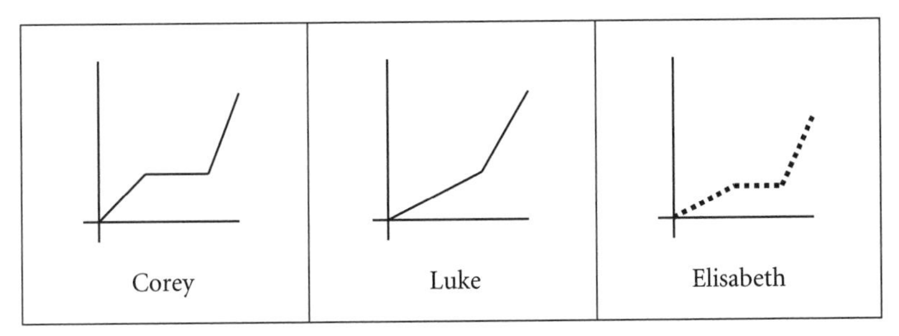

Fig. 13.7. Typical algebra students' responses to the allowance problem

SUMMARY AND IMPLICATIONS FOR TEACHING

In this article, we presented three examples of translation processes that require meaningful, contextual interpretation of representations in a problem-based approach. Providing students with experiences in graphical interpretation, verbal pattern description, qualitative graphical construction, and data generation fosters students' understanding of essential mathematical concepts such as covariation and rate of change. Indeed, the representations become an intellectual tool for students' understanding.

Students' abilities to interpret representations of linear functions vary according to the nature of the given patterns of covariation (Coulombe 1997). Specifically, positively sloped relations, where both variables are increasing appear to be the easiest for students to interpret. Somewhat more difficult are negatively sloped relations, where the independent variable is increasing but the dependent variable is decreasing. Finally, students seem to have the most difficulty when one variable is increasing and the other variable is neither increasing nor decreasing. Interpretations of representations of these stable situations may serve as tools for students to use to build their mental images of constant patterns and functions.

Graphs and tables can be effective tools for investigating linear (and nonlinear) patterns in a variety of contexts. For example, the weight-loss problem discussed in this article allowed students to explore three linear patterns through graphical interpretation and data generation. The weight-loss problem can be extended with alternative representations such as words and symbols. Students might be asked to generate a verbal rule for each "third" of the graph and then to translate each of those rules into symbols.

Describing the relationship between the independent variable and the dependent variable is an important building block in developing students' understanding of rate of change. Consider that in rate-of-change problems, time is usually the independent variable (x) and the variable that is changing over time is the dependent variable (y). Initially, students can use verbal communication to describe the rates. For example, "as the years increase, the population increases," or "as the years increase, the capacity of the landfill decreases." Teachers can ask students to describe these relationships on a graph without a scale. In the first relationship, the slope is positive; the second relationship has a negative slope. Then the students can generate data tables that reflect the relationships between the variables. Finally, students can relate their ideas of change to numbers and symbols. For example, if we let p represent the population increase (the dependent variable) and x represent the number of years (the independent variable), then the relationship might be written as $p = 25\,000x + 350\,000$.

Another classroom strategy is to ask students to consider the context of rate-of-change problems. Just as elementary school teachers ask students to write story problems for the multiplication and division of whole numbers, algebra teachers can assess their students' understanding of algebraic representations by asking them to write word problems from tables or graphs. The independent variable for these representations is time, and students are asked to define a dependent variable that changes over time in a manner defined by the representation. For example, given a graph of $y = 2x$, where x represents time, students are asked the following:

> Write a problem scenario for this positively sloped graph where time is the independent variable. What kind of scale could you put on this graph to fit your problem?

Fluency with multiple representations of mathematical relationships plays a significant role in the successful development of algebraic thinking. Interpretation and translation can also be useful tools for teachers' ongoing assessment of how students are conceptualizing patterns and functions. Additionally, representations can serve as useful guides for instructional planning by helping teachers to see more clearly what students understand and which mathematical ideas are still developing. As teachers listen carefully to students' ideas, they can help students connect their personal representations to more conventional ones such as numerical tables, Cartesian graphs, and algebraic symbols.

REFERENCES

Coulombe, Wendy N. "First Year Algebra Students' Thinking about Covariation." Ph.D. diss., North Carolina State University, 1997.

Janvier, Claude. "Translations Processes in Mathematics Education." In *Problems of Representation in the Teaching and Learning of Mathematics,* edited by Claude Janvier, pp. 27–32. Hillsdale, N.J.: Lawrence Erlbaum Associates, 1987.

Moschkovich, Judit, Alan Schoenfeld, and A. A. H. Arcavi. "Aspects of Understanding: On Multiple Perspectives and Representations of Linear Relations and Connections among Them." In *Integrating Research on the Graphical Representation of Functions,* edited by Thomas A. Romberg, Elizabeth Fennema, and Thomas P. Carpenter, pp. 69–100. Hillsdale, N.J.: Lawrence Erlbaum Associates, 1993.

National Council of Teachers of Mathematics (NCTM). *Principles and Standards for School Mathematics.* Reston, Va.: NCTM, 2000.

14

Promoting Multiple Representations in Algebra

Alex Friedlander

Michal Tabach

MANY teachers and researchers know that the presentation of algebra almost exclusively as the study of expressions and equations can pose serious obstacles in the process of effective and meaningful learning (Kieran 1992). As a result, mathematics educators recommend that students use various representations from the very beginning of learning algebra (National Council of Teachers of Mathematics [NCTM] 2000).

The use of verbal, numerical, graphical, and algebraic representations has the potential of making the process of learning algebra meaningful and effective. In order that this potential be realized in practice, we must be aware of both the advantages and disadvantages of each representation:

- *The verbal representation* is usually used in posing a problem and is needed in the final interpretation of the results obtained in the solution process. The verbal presentation of a problem creates a natural environment for understanding its context and for communicating its solution. Verbal reasoning can also be a tool for solving problems and can facilitate the presentation and application of general patterns. It emphasizes the connection between mathematics and other domains of academic and everyday life. But the use of verbal language can also be ambiguous and elicit irrelevant or misleading associations; it is less universal, and its dependence on personal style can be an obstacle in mathematical communication.
- *The numerical representation* is familiar to students at the beginning algebra stage. Numerical approaches offer a convenient and effective bridge to algebra and frequently precede any other representation. The use of numbers is important in acquiring a first understanding of a problem and in investigating particular cases. However, its lack of generality can

be a disadvantage. A numerical approach may not be very effective in providing a general picture; as a result, some important aspects or solutions of a problem may be missed. Thus, its potential as a tool for solving problems may be sometimes quite limited.

- *The graphical representation* is effective in providing a clear picture of a real valued function of a real variable. Graphs are intuitive and particularly appealing to students who like a visual approach. But graphical representation may lack the required accuracy, is influenced by external factors (such as scaling), and frequently presents only a section of the problem's domain or range. Its utility as a mathematical tool varies according to the task at hand.
- *The algebraic representation* is concise, general, and effective in the presentation of patterns and mathematical models. The manipulation of algebraic objects is sometimes the only method of justifying or proving general statements. However, an exclusive use of algebraic symbols (at *any* stage of learning) may blur or obstruct the mathematical meaning or nature of the represented objects and cause difficulties in some students' interpretation of their results.

The importance of working with various representations is a result of these and other advantages and disadvantages of each representation and of the need to cater to students' individual styles of thinking. Thus, both curriculum developers and teachers should be aware of the need to work in an environment of multiple representations—that is, an environment that allows the representation of a problem and its solution in several ways (usually some or all of the four representations mentioned above). Although each representation has its disadvantages, their combined use can cancel out the disadvantages and prove to be an effective tool (Kaput 1992). Similarly, the Representation Standard for grades 6–8 in the new *Principles and Standards for School Mathematics* relates to the solution of algebraic problems in general and of situations based on linear functions in particular by addressing the following recommendation (NCTM 2000, p. 281):

> Students will be better able to solve a range of algebra problems if they can move easily from one type of representation to another. In the middle grades, students often begin with tables of numerical data to examine a pattern underlying a linear function, but they should also learn to represent those data in the form of a graph or equation when they wish to characterize the generalized linear relationship. Students should also become flexible in recognizing equivalent forms of linear equations and expressions. This flexibility can emerge as students gain experience with multiple ways of representing a contextualized problem.

More specifically, Ainsworth, Bibby, and Wood (1998) mention three ways that multiple representations may promote learning: (*a*) it is highly probable that different representations express different aspects more clearly and that, hence, the information gained from combining representations will be greater than what can be gained from a single representation; (*b*) multiple representations constrain each other, so that the space of permissible operators becomes smaller; (*c*) when required to relate multiple representations to each other, the learner has to engage in activities that promote understanding.

With algebra learning, the use of computers contributes considerably to the promotion of multiple representations (Heid 1995). As students work with spreadsheets and graph plotters, algebraic expressions become a natural requirement and provide an effective means for obtaining a numerical and graphical representation of the relevant data. In a learning environment that lacks computers, drawing graphs or producing extended lists of numbers tends to be tedious and unrewarding.

In the process of solving a problem, isolating representations can be difficult. Thus in most situations, any approach is accompanied by verbal explanations or by numerical computations. In this article we restrict ourselves to the use of representations as *mathematical tools* (and less as means of communication) in the context of beginning algebra. The use of a sequence or a table to answer a question will be an example of a numerical approach, whereas the use of verbal reasoning (possibly including some computations and numbers) will be considered a verbal approach. The use of graphs or algebraic expressions is easier to define and detect.

We cannot expect the ability to work with a variety of representations to develop spontaneously. Therefore, when students are learning algebra in either a technologically based or a conventional environment, their awareness of and ability to use various representations must be promoted actively and systematically. We describe some ways in which tasks can be designed to promote the use of multiple representations. The following section presents, as an example, an activity taken from a beginning algebra course for seventh-grade students and discusses its potential to achieve this goal. We also report some findings about students' use of representations in an assessment task given at the end of one week of work on the activity.

DESIGNING TASKS

In our attempts to promote student thinking and actions in a variety of representations, we found some effective types of tasks and questions. Our analysis of the structure of an activity called *Savings* illustrates the claim that tasks can be designed to encourage the simultaneous use of several representations.

Describing the Problem Situation

Questions, tasks, or even more complex activities are usually presented in one representation, and they may, or may not, require the solver to make a transition to another representation. For example, students may decide to solve a verbally posed problem graphically or algebraically. (In some classrooms, the use of verbal reasoning to solve an algebra problem has not yet received full legitimization.)

We found that presenting various parts of a problem situation in different representations encourages flexibility in students' choice of representations in their solution path and increases their awareness of their solution style. The presentation of a problem in several representations gives legitimization to their use in the solution process. Moreover, to understand and solve such a problem, most students perform frequent transitions between representations and perceive them as a natural need rather than as an arbitrary requirement. Posing a problem in several representations is particularly suitable for situations that require the parallel investigation of several methods, quantities, and so on. Figure 14.1 presents the *Savings* problem situation. In this activity, students investigate the weekly changes in the savings of four children, where the savings of each child is presented in a different representation.

The savings of Dina, Yonni, Moshon, and Danny changed during the last year, as described below. The numbers indicate amounts of money (in dollars) at the end of each week

Dina: The table shows how much money Dina had saved at the end of each week. The table continues in the same way for the rest of the year.

Week	1	2	3	4	5	6	7	8	9 ...
Amount	7	14	21	28	35	42	49	56	63 ...

Yonni: Yonni kept his savings at $300 throughout the year.

Moshon: The graph describes Moshon's savings at the end of each of the first 20 weeks. The graph continues in the same way for the rest of the year.

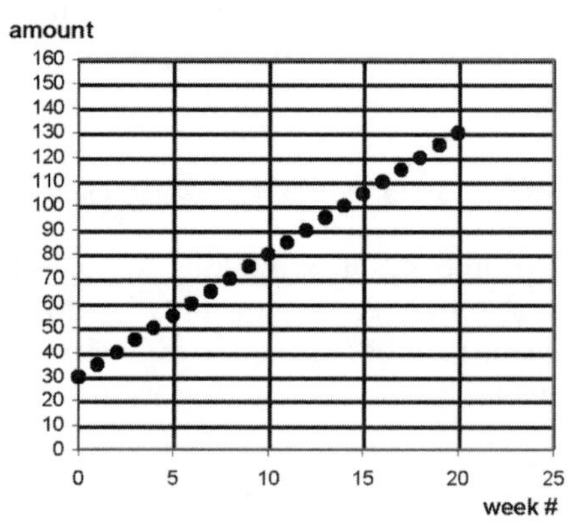

Danny: Danny's savings can be described by the expression $300 - 5x$, where x stands for the number of weeks.

Fig. 14.1. *Savings*—problem situation

Posing Investigative Questions

The presentation of the problem situation is followed by a variety of questions aimed at leading students through their investigations. These questions are posed for a variety of reasons. The following categories of tasks are examples of ways to design activities that relate to our agenda of multiple representations. We illustrate each category by a sample of questions from the *Savings* activity.

Getting acquainted with the initial representation

The first questions require students to analyze each component in its original presentation and make some extrapolations or draw some conclusions. At this stage many students avoid any transitions from one representation to another. In our example, we posed questions about the savings of each child. First, we asked for the amount of money at the end of a week that was specifically included in the data. We then asked students to extrapolate to a week not represented. Finally, we asked for the week corresponding to a given amount of money.

Explicit requests for transitions between representations

At the next stage, we require students to work in a specific representation. The following two activities illustrate this stage.

- Describe in words how the savings of each child changes through the year.
- Given the graphs of the savings of all four children throughout the year, identify each graph and find the meaning and the value of each intersection point.

Exploratory questions

Finally, we ask students more complex and open-ended questions. At this stage, we expect them to choose their own method of representation and solution path. In our activity, we asked the students to compare the savings of the four children.

- Compare the savings of two out of the four children.

  Use words like "the savings increase (or decrease),"
  "the savings increase or decrease at a rate of ...,"
  "who has a larger (or smaller) amount at the beginning (or end)," and
  "larger (or smaller) by ... , double ... , equal."

  Use tables, graphs, expressions, and explanations.

- Add another child to your comparison.

Posing Reflective Questions

Reflection has several important aspects. It helps students become aware of the possiblity of using various representations, exposes them to the advantages and disadvantages of these possibilities, and acquaints them with various ways of presenting the solution to a problem. Reflective questions allow students to distance themselves from actions undertaken previously and hence lead them to evaluate their own and others' actions. Moreover, the ability to reflect on the solution of a problem increases considerably the solver's mathematical power (Hershkowitz and Schwarz 1999). As with multiple representations, we cannot rely exclusively on a spontaneous development of the ability to reflect. The following types of tasks are examples of ways to design activities that make reflection an integral part of the solution process.

Description of work

The requirement to describe one's work is attached to many questions. This "habit" is more than routine, and its importance is beyond the need to document the solution. It allows students to reevaluate their solution strategies and eventually to consider other possibilities. Sometimes we make the task more specific by attaching to the text of a problem a blank page called *Work Area,* with the words *Tables, Graphs, Expressions,* and *Descriptions* on various parts of the page. In this example, the use of any particular representation is recommended, but optional. At other times, we directly ask students to mention the representation they used on each occasion.

Commenting on others' work

Presenting the work of one or several (fictional) students reduces the burden of getting involved in the actual process of solving a problem and allows students to relate to, and reflect on, particular aspects of the solution. Here is an example:

Ran wanted to find how much Dina had saved by the end of the 15th week. Vered suggested continuing the table a little more.

Week	11	12	13	14	15
Amount	77	84	91	98	105

She looked at the table and found that the amount is $105. Motty claimed that he had another way. Since Dina had no savings at the beginning of the year and her savings increased by 7 each week, she would have 7 times the number of weeks—that is, $7 \cdot 15 = 105$. Do you think that both methods are correct? Which method do you prefer?

Asking students to design their own questions

Another possible way to raise awareness of the potential of various representations is to give a problem situation in one or multiple representations (possibly collected from students' previous work) and ask students to design (and solve) a question that in their view can be answered by using the given representation. In our activity, for example, we can pose this task and enclose the tables, graphs, expressions, or verbal descriptions of the savings of *all* four children.

Asking for reflection on mathematical concepts

Journal items are particularly appropriate for asking students to reflect on possible ways to answer the posed questions and to describe their solution. Thus, toward the end of our activity, we require students to construct a concept map on ways to represent data and solutions. They were also encouraged to discuss the advantages and disadvantages of using a particular representation.

Allowing time for reflection

The solution of complex problems over a longer period of time (in our situation, five lessons spread through one week) creates, of itself, further opportunities for spontaneous or induced reflection.

In the next section, we consider the use of various representations in the solution of a task by two classes of beginning algebra students, who worked on *Savings* and other similar activities.

ASSESSING STUDENTS' USE OF REPRESENTATIONS

After a week of investigating the *Savings* activity (including one lesson of work with Excel), the teachers of two seventh-grade beginning algebra classes gave an assessment task related to the same context. The task was given about two months after the beginning of the course to seventy students who worked in pairs and without computers. Although the assessment of the students' work had a wider scope, we present here only some findings that relate to their use of representations. At the initial stage of the task, the savings of two children during a year were described in a table and a graph. Then, the students were required to answer a sequence of questions and were specifically instructed, both orally and in writing, to show their work and to mention the representation they used in each answer. Figure 14.2 presents the first seven (of ten) questions in this task.

The table and the graph below describe the savings of Danny and Moshon during the year.

Week #	Savings		Week #	Savings	
	Danny's	Moshon's		Danny's	Moshon's
0	300	30	26	170	160
1	295	35	27	165	165
2	290	40	28	160	170
3	285	45	29	155	175
4	280	50	30	150	180
5	275	55	31	145	185
6	270	60	32	140	190
7	265	65	33	135	195
8	260	70	34	130	200
9	255	75	35	125	205
10	250	80	36	120	210
11	245	85	37	115	215
12	240	90	38	110	220
13	235	95	39	105	225
14	230	100	40	100	230
15	225	105	41	95	235
16	220	110	42	90	240
17	215	115	43	85	245
18	210	120	44	80	250
19	205	125	45	75	255
20	200	130	46	70	260
21	195	135	47	65	265
22	190	140	48	60	270
23	185	145	49	55	275
24	180	150	50	50	280
25	175	155	51	45	285
			52	40	290

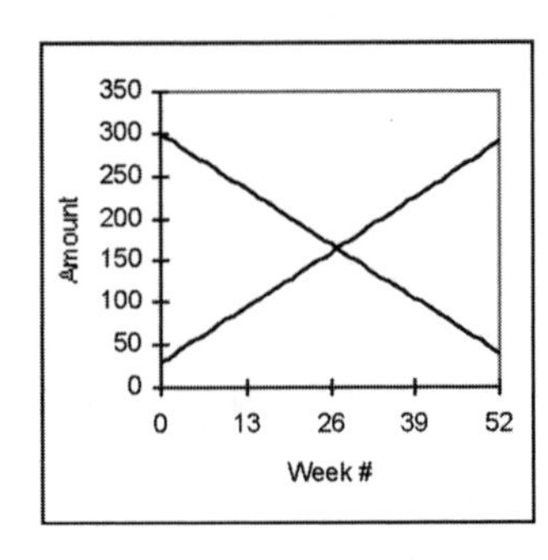

Fig. 14.2. *Savings*—assessment task data

Describe the work that you do to answer the questions below.

Please describe in detail all your (right or wrong) attempts, and the representations (table/expressions/graph/words) that you use to answer each question.

Important Remark:
The solution process is more important than the final result.
A detailed description of your work will improve your assessment.

The questions:

1. How much had Moshon saved after half a year? And how much did Danny have at the same time?
2. After how many weeks did each of the two children have $210?
3. When was the difference between their savings $60? In whose favor was the difference?
4 Find the week with the largest difference between their savings.
5. Find the week when their savings were equal.
6. Find the week when the savings of one were double that of the other. In whose favor?
7. Danny and Moshon decided to pool their savings in order to buy a $400 walkie-talkie. Find the week in which they can realize their intention.

Fig. 14.2 (*cont.*) The first seven questions

The teachers mentioned algebraic expressions at the beginning of the task as one of the four possible representations but did not actually give any. Our impression from the students' classroom work indicated that at this stage of the course, they preferred numerical or verbal solutions and made a more limited use of graphs, and even less use of algebraic expressions. We wanted, however, to have a more detailed picture of students' preferences and flexibility in their choice of representations. According to our (expert) view, different questions in the task favor different representations. Thus, the first two questions clearly favor the use of the given table of numbers, whereas to find the largest difference between their savings (question 4), the use of graphs is more advantageous. In our opinion, the other questions could be answered with a reasonable investment of effort by choosing from several possible representations.

As expected, the use of the table of numbers was dominant. However, each question attracted various representations. To find the savings after half a year (question 1), the majority of students made direct use of the given table, as in the following solution:

> - We looked at Danny and Moshon's table and found the savings for the 26th week because $52 \div 2 = 26$.
>
>   52 → weeks in a year
>   2 → divided by 2 for half a year
>
> Moshon's savings in the 26th week are $160 and Danny's savings are $170.

Using the table to answer question 1 seems natural and simple. The following two examples show, however, that the approaches to the solution were quite varied. Some pairs of students preferred the algebraic expressions (first example) or graphs (second example).

> - After half a year, Moshon had $160. During the first week, Moshon had $30, and each week he added $5. Therefore we made the expression $30 + 5x$. x is the number of weeks and in order to compute [the amount after] half a year, the expression will be $30 + 5 \cdot 26$ and in order to calculate on a calculator we need [to keep] the order of operation. Thus, we found $5 \cdot 26 = 130$ and added $30 = 160$.
> - We marked the midpoint of the horizontal axis and drew a line upwards (until our line intersects Danny's and Moshon's points). From Danny's and Moshon's points we drew a horizontal line to the vertical axis. We discovered that Danny had $175 by the middle of the year and Moshon had only $150.

Question 7 related to the possibility of Danny's and Moshon's reaching a common sum of $400 and attracted many verbal solutions, like the not completely correct reasoning in the following example.

> - Danny and Moshon will never get the walkie-talkie because when Danny will have $300 (his largest amount) Moshon will have only $30 and when Moshon will have $290 (his largest amount) Danny will have only $40 and therefore the largest amount that they can reach is $330.

An analysis of the students' answers to all seven questions showed that students were remarkably flexible in their use of representations. Only five pairs of students were consistently numerical. All the others used two, three, or four representations (31, 37, and 12 percent, respectively). Sometimes, a pair used more than one representation to answer a question. Such transfers between representations usually occurred when work with the initially chosen representation seemed too difficult or unrewarding. The following two answers to question 3 (finding cases when the difference between the savings is $60) illustrate this situation.

- We compared the expressions: Moshon 30 + 5*x* and Danny 300 – 5*x* and then ... ah ... then we switched to an unsystematic and silly search of each number in the table (the results: the 21st, and the 33rd week).

In this example, the pair of students attempted to answer the question algebraically, but a lack of knowledge about how to use this tool forced them to switch to the numerical approach, in spite of their awareness of its disadvantages. In the next example, another pair makes a transition from a numerical to a graphical representation. The students make the transition to graphs when their initial attempt provides an incomplete answer.

- On the 21st week Danny had $60 more than Moshon. On the 33rd week Moshon had $60 more than Danny did. We looked for a difference of less than $100 and when we found them, we looked in Moshon's column [to find] when do we have to add or subtract $60 to get Danny's amount in the same week. We found in Danny's column $195 and in the same week we found there in Moshon's column $135. Then we saw in the graph that the same case happens, only that [the amount of] Moshon is larger by $60 than Danny's amount.

Table 14.1 presents the distribution of the students' choice of representations on each of the first seven questions. Besides the obvious dominance of the numerical representation, it should be noted that some questions attracted a relatively large proportion of other representations. Thus, about 20 percent of the answers to question 5 (finding when the savings are equal) were based on graphs, and more than half of the answers to question 7 (finding when the total savings exceed $400) were either verbal or algebraic.

CONCLUSION

Many mathematics educators recommend using multiple representations in algebra. We have tried to illustrate some concrete ways of enhancing students' awareness of these advantages and their ability to use them in their routine work. The design of the *Savings* activity helped us illustrate our belief that the promotion of multiple representations depends in the first place on the presentation of a problem situation and on the nature of the questions asked. These should suggest, legitimize, recommend, and some

TABLE 14.1
Choice of Representation as a Percent of Total Responses for Each Question and the Assessment Task as a Whole*

Question	Numerical	Verbal	Graphical	Algebraic	Unidentifiable
1	68	0	5	16	11
2	74	3	5	13	5
3	71	10	7	7	5
4	51	14	19	0	16
5	61	5	24	0	10
6	59	13	13	2	13
7	25	41	8	13	13
Total	60	13	12	7	8

*The total number of students was 70 (35 pairs). However, if two representations were used in a pair's answer to a question, each representation was counted separately in the corresponding column.

times even require more than one representation. To internalize the principle of multiple representations, student reflection on these actions is also needed and should be promoted by the task design.

The *Savings* activity and its follow-up task were conducted as regular classroom activities and were not planned and carried out as research. However, the analysis of students' responses supports our claim that suitable problem posing and questioning—and systematic encouragement of students' experimentation with various representations—can increase the awareness of, and the ability to use, various representations in the solution of a problem.

The predominant use of the numerical representation was expected. We relate this preference to the students' early stage in their learning of algebra and to the fact that in many situations the nature of a task makes the use of a numerical approach mathematically sound.

The good news, however, is that if students are given an appropriate learning environment, they will be able and willing to employ a wide variety of solution tools and paths. In our analysis of students' work, we found that the choice of a representation can be the result of the task's nature, personal preference, the problem solver's thinking style, or attempts to overcome difficulties encountered during the use of another representation. Frequently, the choice of representation is influenced by a combination of several factors. To answer a question, students may choose a representation on the basis of their analysis of the problem and personal preference, and they may switch to another representation at a later stage as a result of difficulties in the solution process.

REFERENCES

Ainsworth, Shaaron E., Peter A. Bibby, and David J. Wood. "Analysing the Costs and Benefits of Multi-Representational Learning Environments." In *Learning with Multiple Representations,* edited by Maarten W. van Someren, Peter Reimann, Henry P. A. Boshuizen, and Ton de Jong, pp. 120–34. Oxford, U.K.: Elsevier Science, 1998.

Heid, M. Kathleen. *Algebra in a Technological World. Curriculum and Evaluation Standards for School Mathematics* Addenda Series, Grades 9–12. Reston, Va.: National Council of Teachers of Mathematics, 1995.

Hershkowitz, Rina, and Baruch Schwarz. "Reflective Processes in a Mathematics Classroom with a Rich Learning Environment." *Cognition and Instruction* 17 (1999): 65–91.

Kaput, James J. "Technology and Mathematics Education." In *Handbook of Research on Mathematics Teaching and Learning,* edited by Douglas A. Grouws, pp. 515–56. New York: Macmillan, 1992.

Kieran, Carolyn. "The Learning and Teaching of School Algebra." In *Handbook of Research on Mathematics Teaching and Learning,* edited by Douglas A. Grouws, pp. 390–419. New York: Macmillan, 1992.

National Council of Teachers of Mathematics (NCTM). *Principles and Standards for School Mathematcs.* Reston, Va.: NCTM, 2000.

15

From Fibonacci Numbers to Fractals

Recursive Patterns and Subscript Notation

Deborah S. Franzblau

Lisa B. Warner

WHEN one of us taught the college course Mathematics for Liberal Arts, several students were in a panic after the first class: "We enjoyed the game we played tonight—but we're not going to do any algebra, are we?" "You just did!" replied the instructor. The students had just played a number game in which they had to formulate a winning strategy. "That was algebra?" one student asked incredulously. Algebraic thinking did not frighten these students—but some other aspect of algebra did.

On the second evening, the class tried an activity used by the instructor to teach algebra to middle school students: What's My Rule? Using the following table, students had to describe a rule relating the input N and the answer A, then try to give a formula for the rule.

N	A
4	17
1	2
7	50
2	5
9	82

This work is supported in part by the NSF NY Collaborative for Excellence in Teacher Preparation under grant # DUE 945 3606. All opinions expressed are those of the authors and not necessarily those of the Collaborative.

A few students had trouble describing a rule: "I'm coming up with the answers, but I can't explain what I did in my head," one lamented. But the majority seemed to agree with the student who said, "I'm really good at explaining the function in words, but I get confused with formulas—can't I just stop with the words?"

Later in the course, the students were introduced to fractals and Sierpinski's triangle; the first three stages of the construction are shown in figure 15.1. After a discussion, we looked at related numerical sequences, such as the number of triangles shaded at each stage, shown in the following chart:

Stage	0	1	2	3	4	...	N
Shaded triangles	1	3	9	27	?	...	?

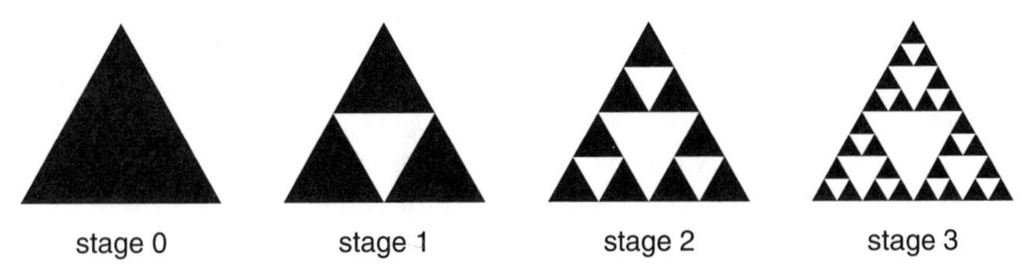

Fig. 15.1. Construction of the Sierpinski triangle, stages 0–3.

The students quickly saw that each new term can be found by multiplying the preceding term by 3. One student explained why: Each stage contains three smaller copies of the preceding stage. When it came to writing down a formula, however, many of the students became increasingly frustrated.

This story illustrates a central theme in this article: Students can recognize and continue patterns readily—what gives them trouble is recording their thinking with symbols.

Both the sequence of stages of Sierpinski's triangle and the list of the numbers of shaded triangles (1, 3, 9, 27, ...) are examples of *recursive patterns*, sequences in which each object can be generated systematically from the preceding objects. Recursive patterns are often easy to understand but difficult to record.

Sequence Notation: Recursive and Explicit Formulas

There are two standard mathematical representations for sequences, *subscript notation* and *functional notation*. The first three terms of the sequence (1, 3, 9, 27, ...) can be represented in subscript notation as $t_1 = 1$, $t_2 = 3$, $t_3 = 9$ and in functional notation as $t(1) = 1$, $t(2) = 3$, $t(3) = 9$. In each example, t_n or $t(n)$ is the term of sequence t in position n, or, more abstractly, the value of function t evaluated at n. The choice of the index set $\{n\}$ is a matter of taste or convenience; for example, the first three terms can also be written as $t_0 = 1$, $t_1 = 3$, $t_2 = 9$.

A recursive formula for a sequence, called a *recurrence relation* or a *difference equation,* is a formula in which each term in the sequence is a function of one or more previous terms. A recursive formula for the sequence (1, 3, 9, 27, ...) is $t_n = 3t_{n-1}$ (in subscript notation) or $t(n) = 3t(n-1)$ (in functional notation). To represent a sequence uniquely, we also need an appropriate set of *initial values;* in this example, only one initial value is needed: $t_1 = 1$. The recurrence is then valid for $n \geq 2$. The labeling of terms is a matter of choice; for example, an equivalent recursive formula is $t_{n+1} = 3t_n$.

The *Fibonacci sequence* (1, 1, 2, 3, 5, 8, 13, ...) is a famous recursive number pattern, in which each term is the sum of the two previous terms. A recursive formula is $F_n = F_{n-1} + F_{n-2}$. Given initial values $F_1 = 1$, $F_2 = 1$, the recurrence is valid for $n \geq 3$.

In an *explicit formula,* each term is obtained directly from the independent variable(s). For example, (1, 3, 9, 27, ...) has the simple explicit formula $t_n = 3^{n-1}$ $(n \geq 1)$. Explicit formulas for the Fibonacci sequence are more complicated (see Epp [1995, p. 460]).

What Is the Value of Sequence Notation?

> *When I use "the previous result" and arrows on the board, all is fine, but that notation breaks down quickly.* (Charles H., middle school teacher)

> *If introduced after conceptual understanding is established, then the subscript notation can be very useful, both as a means for implementing the recursive algorithm and as a way to unify different topics.* (Hart 1998, p. 266)

The recent *Principles and Standards for School Mathematics* (National Council of Teachers of Mathematics [NCTM] 2000) recommends in the Algebra Standard (pp. 296) that students learn to use standards sequence notation during grades 9–12 for recursive as well as explicit relationships. For students who later study discrete mathematics, calculus, or computer science, understanding both subscript and functional notation is essential, for example, to work with summations of functions. Also, subscript notation can be used to focus on the values of the terms, whereas functional notation can be used to focus on the method for generating terms. Both notations are useful for computation. Spreadsheets and some calculators use a system equivalent to subscript notation. Computer programming languages often use functional notation.

Many teachers start with simplified variables such as ANSWER and PREVIOUS, or NEXT and NOW (Hart 1998, p. 262).Such notation works well for arithmetic and geometric sequences, but standard sequence notation is more appropriate for the Fibonacci sequence (1, 1, 2, 3, 5, 8, ...) and is in general more flexible. For the triangular numbers (1, 3, 6, 10, 15, 21, ...) it is necessary to introduce another variable: ANSWER = PREVIOUS + STAGE. Sequence notation can reveal the relationships between variables: letting T_n be the nth triangular number, $T_n = T_{n-1} + n$. Moreover, standard notation connects recursive and explicit formulas. For example, an explicit formula for the triangular numbers can be written as $T_n = n(n + 1)/2$.

DECIDING TO TEACH SUBSCRIPT NOTATION

In the past, we tried to avoid sequence notation when possible, agreeing, for example, with Hart (1998, p. 265) that "formal subscript notation is powerful but dangerous. If subscripts are introduced too soon, students may get bogged down in the technical formalism and miss the basic idea of recursion." Instead we relied on tables, words, or simplified notation, as in Maletsky (1997) or Billstein, Libeskind, and Lott (1997, ch. 1–2). We were dissatisfied with these approaches, however, and concerned that we were shortchanging our students.

Our discussions motivated us to teach students in the Mathematics for Liberal Arts class to use subscript notation. Although the students had difficulties at first, after substantial time for practice and discussion, they were able to use the notation successfully and were gratified by their success. Emboldened by this experience, and knowing the difficulties that more advanced students in calculus or discrete mathematics have with sequence notation, we wondered whether students would benefit from exposure to the notation much earlier in the curriculum. We decided to introduce subscript notation in three seventh-grade mathematics classes: two standard heterogeneous classes and an honors class. We looked carefully at students' written work and interviewed selected students. We hoped to better understand students' thinking and improve our methods for teaching the notation.

Subscript Notation versus Functional Notation

Our decision to focus on subscript notation grew out of a discussion in a class for preservice elementary school teachers. The students, who had seen parentheses only in the context of multiplication, explained that they interpret an expression like $f(5) = 10$ as $f \times 5 = 10$. Even worse, they have an almost irresistable urge to solve for f to obtain $f = 2$.

Also, most middle school students in our district are just learning to use parentheses in multiplication. We believe that introducing a very different use of parentheses at the same time would confuse students unnecessarily.

INTRODUCING STUDENTS TO SUBSCRIPT NOTATION

In the standard seventh-grade classes, we wrote the terms (1, 2, 4, 8, ...), then asked students to look for a pattern, write a description in words, and try to give a formula. No instructions were given on how to create the formula. We interviewed students on audiotape as they worked and then collected their written work. As expected, most students had little trouble recognizing and continuing the pattern. And, in each class, at least two-thirds of the students gave a reasonable written description of their procedure:

> You have to multiply every number by 2 to get your answer.
>
> The way you see your pattern is you double your numbers.
>
> My strategy was that I just doubled the numbers as I went down the chart.
>
> You have to double the number to get the next number until you are done.

The students already had experience with variables and formulas from playing What's My Rule? and from other classroom activities. About a third of the students gave variants of the formulas $N \times 2 = A$, or $N + N = A$, where A meant "the answer" and N meant "the answer before."

As students presented their formulas, using their own notation, they gave descriptions like "you take the answer and multiply by 2." The question "*Which* answer do you mean?" generated such responses as "the answer before" or "the answer you need" as well as "the previous answer." One student, Mike, pointed out that the numbers were powers of 2 ($2^0, 2^1, 2^2, \ldots$). Jessica suggested 2^N as a formula to record this pattern. The students then had two formulas for the same sequence, in which N had a different meaning in each. This result presented a good context in which to introduce subscript notation:

> Now I'm going to show you a more sophisticated way to record your thinking. Let's use n to represent the stage number and F sub n to represent the answer.

One teacher wrote the following on a transparency:

$$n = \text{Stage Number}$$
$$F_n = \text{Answer}$$

Stage: n	0	1	2	3	...
Answer: F_n	1	2	4	8	...
	F_0	F_1	F_2	F_3	...

The teacher then said, "I'm also going to use F sub n minus 1 to mean the previous answer" and added a line to the transparency:

$$F_{n-1} = \text{Previous Answer}$$

Students were then asked a series of questions to help them understand the notation:

> "If n is the stage number, how can we represent the previous stage number?"
>
> "What symbols should we use for the answer *before* the previous answer?"
>
> "If we are looking at Stage 9, what would F sub n be?"
>
> (In Stage 9) "What would F sub n minus 1 be?"
>
> "If the answer, F sub n, is 32, what can we call 16?"

After the students seemed comfortable with the new symbols, they wrote a journal entry on notation for sequences and recursive formulas. In each of the seventh-grade (nonhonors) classes, about a fifth of the students gave a correct formula by using subscript notation: either $F_{n-1} + F_{n-1} = F_n$ or $F_{n-1} \times 2 = F_n$.

Many of the college students, as well as the seventh graders, at first thought that F_{n-1} meant "take the answer and subtract 1," or $F_n - 1$, rather than "the previous answer." Students had to discuss the difference between the two interpretations several times to clarify the meaning.

Some students at first resisted using both F and n. Angela wrote both

$$\underset{n-1}{F} + \underset{n-1}{F} = \underset{n}{F}$$

and $F \times 2 = F$. Johanna wrote $F + F = F$ and $(F - 1) \times 2 = F$. They must have felt like the college student who asked, "Why do you need *all of that* just to mean previous answer? Why don't you just have one F?!"

Probing Further into Student Understanding

The seventh-grade honors class first saw subscript notation earlier that term when studying the Fibonacci numbers (1, 1, 2, 3, 5, 8, ...). (This is the source of the letter F in the formulas above.) Later, as part of a geometry lesson, students generated the following table relating the numbers of vertices, edges, and faces of n-sided prisms.

n = number of sides	3	4	5	6	...
Number of vertices	6	8	10	12	...
Number of edges	9	12	15	18	...
Number of faces	5	6	7	8	...

After reviewing subscript notation and formulas, students wrote journal entries discussing both recursive and explicit formulas for the patterns they found. Most students gave correct formulas, for example, $3n = e$ as an explicit formula for the number of edges e and $F_{n-1} + 3 = F_n$ as a recursive formula with F_n representing the number of edges.

The students' consistent use of F_n as a variable name in recursive formulas reflects the teacher's decision to postpone the introduction of new variable names to allow students to focus on one new visual element at a time. After seeing the formulas in class, many students started using F_n in explicit formulas as well, writing $3n = F_n$ instead of $3n = e$, for example.

Also, virtually all the students wrote the answer to the right of the = sign, using = to mean "gives" or "produces," as discussed in Kieran (1992, p. 393). The correct grammar is "$F_n = F_{n-1} + 3$," where = means "is defined to be."

Even after students had agreed on the correct way to write subscript notation, several continued to write the notation incorrectly. Michael wrote an ambiguous $Fn - 1 + 2 = Fn$ for the number of vertices. Gina and Kristy wrote $Fn_{-1} + 3 = Fn$ for the number of edges; Rachel wrote $F_n{}^{-1} + 3 = F_n$. We thought at first that the reason was visual inexperience or sloppiness. Some students were interviewed to find out more. Michael's responses seemed to confirm the sloppiness theory.

Interviewer: Suppose I give you $F_n = 14$ and I want F sub n plus one [*writing* F_{n+1}].

Michael: That would be the answer after, F of n plus one would be sixteen [*writing* $F_n + 1 = 16$].

Interviewer: Do you see any difference between the way I'm writing F of n plus one and the way you're writing it? [*Question echoes Michael's use of the functional language* "F of n."]

Michael: [*Correcting his written expression to* $F_{n+1} = 16$.] If you had a bigger n—a regular-sized n—that would mean "*Fn*," but when you have the little n, that means "*F of n*."

However, interviews with Gina and Kristy (who were counting edges rather than vertices) showed that there was conceptual confusion behind some of the subscript position errors.

Interviewer: [*Writing* $F_n = 18$] Does this mean something to you?

Gina: Yes, it means the answer is eighteen.

Interviewer: If I write this [F_{n+1}], what does that mean?

Gina: The answer after it ... no, the answer plus one, eighteen plus one is nineteen.

Interviewer: Which one does it mean, the answer afterward or the answer plus one?

Gina and Kristy: The answer plus one.

Interviewer: What if I write this [F_{n-1}]?

Gina: That would be the answer before.

Interviewer: What if I write [F_{n-2}]?

Gina: That would be twelve, since *Fn* is eighteen. [*Correctly seeing that* $F_{n-2} = (F_n - 3) - 3$]

It seemed strange that Gina and Kristy were able to generalize to moving backward in the sequence, but not to moving forward. Samantha also thought F_{n+1} meant "the answer plus one," even though she wrote her formula correctly as $F_{n-1} + 3 = F_n$.

We began to suspect that the students did not completely understand the logic of the notation; that is, they did not see *n* as an independent variable. To check this, we gave students the triangular number problem, based on the pattern shown in figure 15.2.

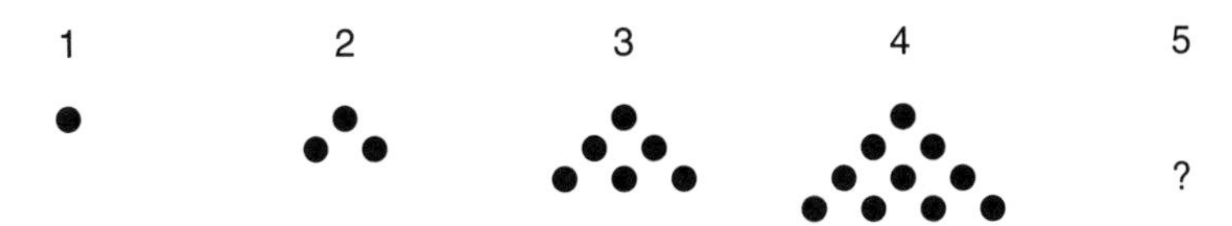

Fig. 15.2. Triangular arrays

Students helped create the following table and were again asked to describe a pattern and give a formula for the pattern.

Number of rows	1	2	3	4	...
Triangular number	1	3	6	10	...

The results confirmed our suspicions: Only four of the thirty-four students wrote their formula as $F_{n-1} + n = F_n$; more than half the students wrote formulas of the form $F_{n-1} + S = F_n$, using *S* to represent the stage number (number of rows) and not seeing that $S = n$.

Samantha wrote $F_{n-1} + N = F_n$. We thought incorrectly that she was not case-sensitive:

Interviewer: Is this (capital) *N* the same as the little *n*?

Samantha: Well, not really because *F* of *n* is the answer but *N* is just representing a number that you're not sure of.

Interviewer: I see two different *N*s here, a big *N* and a little *n*. What does this little *n* represent?

Samantha: The little *n* is part of *F* of *n* and *F* of *n* means "the answer." ... The little *n* doesn't mean anything separately—it's just part of *F* of *n*.

Gina also became confused when asked what the little *n* represented: "It's a variable ... I mean without this it's not like ... you wouldn't know ... but it's not that specific."

Putting together Samantha's and Gina's responses with an earlier statement of Gina's finally cleared up the mystery. When asked what her formula ($Fn_{-1} + 3 = Fn$) meant, Gina said: "The previous answer plus three equals *F* of *n*. *F* of *n* is the answer; *F* of *n* minus one is the answer before; minus one means before it."

This was the source of the confusion: students were viewing F_n as a single symbol for "the answer"; they interpreted $F_{(n-1)}$ as $(F_n)_{-1}$, seeing the −1 as code for "back up one from the answer." With this interpretation, an error such as writing F_n^{-1} is very natural. It is surprising, however, that the same student would not interpret the +1 in F_{n+1} as "move forward one from the answer"; it may arise from having experience only with formulas using F_n, F_{n-1}, and F_{n-2}.

PERSPECTIVES FROM RESEARCH ON SCHOOL ALGEBRA

Research on the teaching and learning of algebra, as described in the survey by Kieran (1992), can shed light on our observations.

The Concept of a Variable

A full understanding of subscript notation such as F_n requires seeing that *F* by itself represents the sequence and that *n* is an independent variable that can take on whole-number values (0, 1, 2, 3, ...). One must recognize that expressions such as $n - 3$ or $n + 1$ are dependent variables representing sequence positions and that F_k is a dependent variable representing the *k*th term of *F*. Even a brief look at the historical development of the concept of a variable, as outlined in Kieran (1992, p. 391), makes clear that substantial time is needed to reach the necessary level of sophistication. Kieran's outline describes a progression of levels of understanding:

- Level 0: using ordinary language to solve problems involving unknowns
- Level 1: using letters as specific unknown numbers
- Level 2: using letters as generalized numbers—(independent) variables that can take on a set of values
- Level 3: using letters as both independent and dependent variables

Moreover, Kieran notes, letters were used to represent unknown values only after Diophantus (c. A.D. 250) and were not used to represent variables before the Renaissance in the 1500s. An explicit distinction between dependent and independent variables required yet another two hundred years.

Most of the students we worked with seemed comfortable with variables at Level 1 or Level 2, in the context of input-output rules, but it is not surprising that they did not see *n* as a variable in its own right in an expression like

$F_{n-1} + 3 = F_n$, since such understanding is at Level 3. The recursive formulas themselves, however, may be useful later as a context for teaching the concept of independent and dependent variables.

Procedural versus Structural Understanding

The central learning model developed in Kieran's survey, based primarily on the work of Sfard, is what Kieran calls the "procedural-structural cycle" (1992, pp. 392–93). In this model, learning any new mathematical concept requires an initial understanding in terms of processes or *procedures* involving familiar mathematical objects, followed by a period of assimilation, and eventually culminating in a new understanding or perspective in terms of a single *structure* or new mathematical object. A similar model appears as part of a theoretical framework used by the Research in Undergraduate Mathematics Education Community, derived from the work of Piaget and Garcia (Asiala et al. 1996). In mathematics, both perspectives are necessary.

For example, when students encountered the sequence (1, 2, 4, 8, 16, ...), they first found a procedure to generate the sequence; either $1 \times 2 = 2$, $2 \times 2 = 4$, $2 \times 4 = 8$, ..., or $1 + 1 = 2$, $2 + 2 = 4$, $4 + 4 = 8$.... A few students saw (procedurally) that $4 = 2 \times 2 = 2^2$, $8 = 2 \times 2 \times 2 = 2^3$, $16 = 2^4$, ..., then realized that $1 = 2^0$ and $2 = 2^1$. Over time, as they meet the same sequence in different contexts, without any calculation, students should recognize it as a geometric sequence with ratio 2, or as the sequence of powers of 2.

In the procedural-structural model of learning, learning new notation can be seen as part of the phase of assimilation needed to learn a new concept. However, our experience suggests that the notation itself is a new mathematical idea, which requires its own separate learning cycle.

Also, from this perspective, it is natural for students to interpret F_n to mean "the answer you want" and F_{n-1} as "back up one to the previous answer" because this corresponds exactly to the recursive procedure they used to compute the terms of the sequence. If students are thinking procedurally, it makes sense that they interpret an expression like $b_n = 2^n$ as "to get the answer you want, keep multiplying 2 by itself until you are done" rather than "to find the term in position n, raise 2 to the nth power." Seeing n as a variable requires a more structural perspective.

TEACHING SEQUENCE NOTATION: RECOMMENDATIONS

Although our findings are preliminary, our observations have helped us clarify some important issues in teaching subscript notation. In this section, we formulate recommendations on the basis of our current understanding.

Calculation Provides Motivation for Learning Notation

Students find recursive patterns both accessible and compelling, whether presented as puzzles or in the context of solving another problem, such as counting shaded regions in the Sierpinski triangle. On their own, even young students often find a way to generate a sequence recursively. For example, when a fourth-grade class was introduced to Fibonacci's classic rabbit problem, "eventually, one of the girls in the class realized that … 'You can get the new number by adding the last two numbers together'"(Kowalczyk 1997, p. 28).

Learning sequence notation, however, requires significant time and effort, and students need to believe that it is worthwhile. The seventh-grade students seemed to find the problem of recording ideas in a formula interesting in its own right, probably because the teacher put a high value on the process and communicated this belief to the students. She also communicated the value of accuracy and clarity, asking, for example, "If a visitor walked into class right now, would she or he be able to use your formula?"

Problems involving calculation, such as finding the thirtieth term of (1, 2, 4, 8, …) or determining when the Sierpinski triangle has more than 500 shaded triangles, provide strong motivation for learning sequence notation, even when the notation is not essential. The college liberal-arts students began to see the value of subscript notation after trying the following version of a classic problem (Phillips et al. 1991, p. 15):

> Miracle Mike makes $1 million per year. A rookie makes only $1 the first year, $2 the second, $4 the third, $8 the fourth, and so on. In year 25, who has earned more money in total?

When students shared and discussed their strategies, it quickly became clear that those who had been able to write down correct formulas, whether explicit or recursive, found it easier to calculate the answer and were less likely to make errors than those who could give verbal descriptions only.

Build Understanding of Variables First

Basic understanding of recursive patterns can be checked by asking students to generate the next few terms of a given pattern and to compute specific terms. The key to student readiness to use sequence notation seems to be the ability to use variables to represent a range of numbers and to create formulas to describe functional rules. Playing games like What's My Rule? and writing both explanations and formulas for rules seem to be effective for teaching these concepts.

Students who are not ready to use subscript notation often simply ignore it at first. Over time, as they see other students using the notation and start to make sense out of it, they start to use it. One useful feature of the notation is that partial use is still mathematically correct. For example, $F_{n-1} + S = F_n$ is a valid formula for the triangular numbers, even if a student does not see that $S = n$. Students can move to a higher level of sophistication when they are ready but still write down a correct formula.

Writing in English Is Valuable

We thought that students who initially created a correct formula without instruction, such as $n \times 2 = a$ for the doubling sequence, would be the first to use subscript notation correctly. To our surprise, this did not occur among either the seventh graders or the college students. Many simply continued to use their original formulas and ignored the new notation. However, almost all the students who did use subscript notation right away had first given a good description of the pattern in words.

This observation suggests that an important first step in learning sequence notation is writing descriptions of both explicit and recursive rules in natural language. An explicit connection between writing descriptions and writing formulas is confirmed in MacGregor (1990).

Jump Right In with Standard Notation

Students who can successfully write down a formula using simpler variables may feel satisfied that they have solved the problem, whereas those who are unable to create a formula believe that subscript notation gives them a needed tool. Our observations also suggest that using simplified variables, such as ANSWER and PREVIOUS or NEXT and NOW, may actually be a barrier to making a transition to standard notation, even though this approach is often recommended (see, for example, NCTM [2000, p. 285]).

Overall, we were satisfied with a direct approach to introducing subscript notation. On the basis of our experience, however, we have started to revise the script. First, we want to show that subscripts can be used in explicit as well as recursive formulas: "If $F_n = 3^n$, what is F_8?" Second, we want to clarify the role of the subscript: "If the stage number n is 7, what is the previous stage number?"

Seeing that many symbols can be used is necessary for building a structural understanding of the notation. When students seem comfortable with subscripts, we want them to know that they can choose their own variable names: "I don't have to use the letters F and n. Many people used s as the stage number, so we could let A sub s (A_s) be the answer. What would the previous stage number be? What would the previous answer be? And how would you write the recursive formula? What if we used k as the stage number and T for the answer?"

However, a question that we find difficult to resolve is how long to use one set of variable names after introducing the notation. Holding on to one set of names allows students to build confidence but misleads them to think that there is only one way to represent a recursive formula. Introducing new names quickly may help students develop flexibility, but it leads to significant initial confusion. We do not know which approach is preferable in the long run.

To avoid confusion with the use of parentheses for multiplication, as discussed above, we taught subscript notation only. If we had students who were already comfortable with functional notation, or who needed it to write computer programs, we would have started with that notation instead. The relative difficulty of learning the two notations and the effect of the order in which they are introduced are important research questions, but they are beyond the scope of our work.

Model Correct Usage but Don't Dwell on Mistakes

Errors in writing subscript notation often arise from logical though incorrect interpretations and not from an inability to distinguish case or position. Other errors, such as misuse of the equal sign or parentheses, seem to arise from students' limited experience or the difficulty of using a familiar symbol in a radically new way.

Full understanding of sequence notation is likely to take several years. Thus, our strategy is to approach errors gradually. The teacher should model the correct use of notation, such as the placement of subscripts and the equal sign, but need not discuss these issues at first.

Time and Patience Are Essential

Students need plenty of time to assimilate this complex notation and to work out their misconceptions. For example, most students need to relearn several times that F_{n-1} means "the previous answer" rather than "the answer minus 1." Peer presentations and class discussion are helpful for correcting errors.

Teachers need to reassure students that although the notation may seem complicated or confusing at first, after practice it becomes more natural and easier to use. Moreover, they should expect students to reach different levels of sophistication depending on their initial cognitive level or background.

Analogies can help reduce student anxiety. For example, a teacher could suggest that students imagine an infinite street where F_n is the number that lives in the house with address n. Or, that they could think of F as an infinite cash-register tape and F_n as the total for customer n. However, in light of the procedural-structural learning model, such analogies are unlikely to help students understand the notation, since they arise from a structural rather than a procedural perspective.

CONCLUSION

Interesting recursive patterns arise naturally in the study of numbers, algebra, probability, and geometry. Such patterns fascinate students and spark creative thinking, so they are quickly becoming staples of the curriculum, as recommended in both the original and the new NCTM *Standards* documents (1989, 2000). For more background on recursive sequences, and discrete mathematics in general, see Kenney (1991), Kenney and Bezuszka (1993), Hart (1998), and Rosenstein, Franzblau, and Roberts (1997).

Before the experiments we have described, we, like many teachers, believed that standard sequence notation is too confusing for students before late high school or college, and we avoided using it. However, it is now clear to us that mastering either subscript or functional notation is a long-term process. It seems essential to introduce it in the curriculum well before students need it in calculus, computer science, probability, or discrete mathematics.

Our experience suggests that middle school is not too early to introduce sequence notation, provided that students already have experience with patterns, verbal descriptions, variables, and formulas. As long as students are already interested in recursive patterns and in devising formulas, the notation, though complex, allows students to express their own thinking in a powerful way.

REFERENCES

Asiala, Mark, Anne Brown, David J. DeVries, Ed Dubinsky, David Mathews, and Karen Thomas. "A Framework for Research and Curriculum Development in Undergraduate Mathematics Education." *Research in Collegiate Mathematics Education* II (1996): 1–32. For an Internet site with a glossary of terms, see www.cs.gsu.edu/~rumec/Papers/glossary.html (February 1, 2001).

Billstein, Rick, Shlomo Libeskind, and Johnny W. Lott. *A Problem Solving Approach to Mathematics for Elementary School Teachers.* 6th ed. Reading, Mass.: Addison-Wesley, 1997.

Epp, Susanna S. *Discrete Mathematics with Applications.* 2nd ed. Boston: PWS Publishing, 1995.

Hart, Eric W. "Algorithmic Problem Solving in Discrete Mathematics." In *The Teaching and Learning of Algorithms in School Mathematics*, 1998 Yearbook of the National Council of Teachers of Mathematics (NCTM), edited by Lorna J. Morrow, pp. 251–67. Reston, Va.: NCTM, 1998.

Kenney, Margaret J., ed. *Discrete Mathematics across the Curriculum, K–12.* 1991 Yearbook of the National Council of Teachers of Mathematics (NCTM). Reston, Va.: NCTM, 1991.

Kenney, Margaret J., and Stanley J. Bezuszka. "Implementing the Discrete Mathematics Standard: Focusing on Recursion." *Mathematics Teacher* 86 (November 1993): 676–80.

Kieran, Carolyn. "The Learning and Teaching of School Algebra." In *Handbook of Research on Mathematics Teaching and Learning,* edited by Douglas A. Grouws, pp. 390–419. New York: Macmillan, 1992.

Kowalczyk, Janice C. "Fibonacci Reflections—It's Elementary!" In *Discrete Mathematics in the Schools,* edited by Joseph G. Rosenstein, Deborah S. Franzblau, and Fred S. Roberts, pp. 25–34. Reston, Va., and Providence, R.I.: National Council of Teachers of Mathematics and American Mathematical Society, 1997.

MacGregor, Mollie. "Writing in Natural Language Helps Students Construct Algebraic Equations." *Mathematics Education Research Journal* 2 (1990): 1–11.

Maletsky, Evan. "Discrete Mathematics Activities for Middle School." In *Discrete Mathematics in the Schools,* edited by Joseph G. Rosenstein, Deborah S. Franzblau, and Fred S. Roberts, pp. 223–36. Reston, Va., and Providence, R.I.: National Council of Teachers of Mathematics and American Mathematical Society, 1997.

National Council of Teachers of Mathematics (NCTM). *Curriculum and Evaluation Standards for School Mathematics.* Reston, Va.: NCTM, 1989.

———. *Principles and Standards for School Mathematics.* Reston, Va.: NCTM, 2000. www.nctm.org/standards/2000/

Phillips, Elizabeth, Theodore Gardella, Constance Kelly, and Jacqueline Stewart. *Patterns and Functions. Curriculum and Evaluation Standards for School Mathematics* Addenda Series, Grades 5–8. Reston, Va.: National Council of Teachers of Mathematics, 1991.

Rosenstein, Joseph G., Deborah S. Franzblau, and Fred S. Roberts, eds. *Discrete Mathematics in the Schools.* Reston, Va., and Providence, R.I.: National Council of Teachers of Mathematics and the American Mathematical Society, 1997.

16

Tracing the Origins and Extensions of Michael's Representation

Regina D. Kiczek

Carolyn A. Maher

Robert Speiser

MICHAEL is a high school student who made perceptive use of a binary number representation to relate two different problems, revealing their similarity of structure. His coding scheme, initially developed to solve a counting problem, eventually enabled him to make deep connections among other problem situations. Michael's work reflects a long-term process in which he and a group of peers explored important combinatorial questions, not by working with single isolated episodes or topic chunks, but rather by revisiting important tasks across a period of years.

In the episodes that follow, which we present as part of a longitudinal study of students' development of mathematical ideas, Michael and his group engaged in a series of combinatorics and probability investigations. The representation that Michael developed to solve one problem and to justify his results became a tool for keeping track of all possibilities in other problem situations. The group members continued to use their own differ-

This work was supported in part by National Science foundation grants #MDR-9053597 (directed by R. B. Davis and C. A. Maher) and REC-9814846 (directed by C. A. Maher) to Rutgers, The State University of New Jersey. Any opinion, findings, conclusions, or recommendations expressed in this publication are those of the authors and do not necessarily reflect the views of the National Science Foundation.

ent representations to list possibilities; still, they were able to explain Michael's representation and relied on it to check their own ideas. As time went on, Michael was able to use his representation to understand and explain the addition rule of Pascal's triangle. We begin the story with Michael's written explanation, given in the unedited e-mail that was sent to one of the researchers (fig. 16.1).

In this e-mail message, Michael indicated that he had been thinking about a question posed some time before. Approximately three months earlier, in a task-based interview, Michael had reviewed how he used binary numbers to represent the eight different pizzas that can be made when three toppings are available. Referring to the sequential list of binary numbers he had written, Michael explained that 000 was the code for a plain cheese pizza with no toppings and 111 was the code for a pizza with "everything on it." When the teacher-researcher inquired about the distribution of those pizzas, Michael brought up Pascal's triangle.

Teacher-researcher: Can you give me the distribution of those kinds of toppings—how many with none, how many with one, how many with two, and how many with three? Can you tell me what that looks like?

Michael: Yeah, I could. I could tell you what ... [*pause*]. Have you ever heard of—oh, you probably have—that triangle?

Teacher-researcher: Pascal's?

Michael: Yeah. That has a lot to do with it.

As Michael wrote the first several rows of the triangle, he stated that the entries corresponded to the number of pizzas in each category. He showed, for example, that the entries 1, 3, 3, 1 in row 3 of Pascal's triangle gave the occurrence of each type of pizza when selecting from three toppings, indicating in his list the three codes for pizzas with one topping and the three codes for those with two toppings. He then extended this explanation to pizzas that could be made when selecting from four toppings, pointing out the groups of pizzas on his list that corresponded to the entries 1, 4, 6, 4, 1 in the next row of the triangle.

Michael knew the addition rule for Pascal's triangle; indeed, he used it to generate the first few rows. The question posed by the teacher-researcher, prompting Michael's e-mail response, was how the metaphor of counting pizzas connected to the addition rule. During the remainder of the interview, Michael provided specific examples to illustrate how the addition of another topping would affect both his representations and the entries in a particular row. He agreed to put his ideas down on paper and send them to us.

Hi this is Mike. Here are my thoughts on the question that you gave me a while ago. When I was asked to come up with a solution to the pizza problem (a long time ago) I used the binary numbers to give an answer.

M=mushoom 1=topping is present
P=peppers 0=topping is not present
S=sausage ()=number expressed by the binary "code"
p=pepperoni

4-topping pizza

MPSp

0 0 0 0 - no toppings (0)
0 0 0 1 – pepperoni (1)
0 0 1 0 – sausage (2)
0 0 1 1 - sausage and pepperoni (3)
0 1 0 0 – peppers (4)
0 1 0 1 - peppers and pepperoni (5)
0 1 1 0 - peppers and sausage (6)
0 1 1 1 - peppers, sausage, and pepperoni (7)
1 0 0 0 – mushrooms (8)
1 0 0 1 - mushrooms and pepperoni (9)
1 0 1 0 - mushrooms and sausage (10)
1 0 1 1 - mushrooms, sausage, and pepperoni (11)
1 1 0 0 - mushrooms and peppers (12)
1 1 0 1 – mushrooms, peppers, and pepperoni (13)
1 1 1 0 – mushrooms, peppers, and sausage (14)
1 1 1 1 – mushrooms, peppers, sausage, and pepperoni (15)

There is a relation ship between that and Pascal's Triangle

1
1 1
1 2 1
1 3 3 1
1 4 6 4 1
1 5 10 10 5 1

You see, in the pizza example above there were

1 pizza without toppings
4 with one topping
6 with two toppings
4 with three toppings
1 with all four

How come these numbers look so familiar???
Look at the fifth row of Pascal's Triangle!

Fig. 16.1. Michael's e-mail message

If you added another topping to the list like anchovies, you would have:

1 with 0 toppings
5 with 1 toppings
10 with 2 toppings
10 with 3 toppings
5 with 4 toppings
1 with 5 toppings

Why?
Here's why.

The way you make the triangle is by taking the numbers from the row up ahead and adding every two together.(think of having zeros on each side of the "triangle")

```
1 3 3 1 -> 0+1,   1+3,   3+3,   3+1,   1+0
            1      4      6      4      1
```

In the 1 3 3 1 sequence the first position represents the binary numbers with all zeros (no topping pizza)

but when you add another topping, it could either have it or not:

```
      0000
     /
000<    or
     \
      0001
```

So now this 1 pizza combination turned into two. So do all the other combinations. When you add a topping you put a one on the end of it if not you put another zero onto it.

This explains why the Pascal's triangle works:

```
             1
          1     1
       1     2     1
    1     3     3     1
 1     4     6     4     1
1   5    10    10    5    1
```

The underlined 1 represents a pizza with no toppings with a choice of 2. The 2 represents 2 pizzas with 1

Fig. 16.1 (*cont.*)

topping with a choice of 2 toppings and the 3 represents 3 pizzas with one topping with a choice of 3.

Why do we add?

We add because of the fact that every combination will get another place (a 1 and a 0) therefore it doubles the amount of combinations. The reasons why we add numbers that are next to each other is simply because the underlined 1 will become into two new pizzas after another topping choice is added. One of those will be the same (no toppings) and the other will have one topping. The same will happen for the 2. The two will become 4. 2 the same (one topping) and two with two toppings. Now you have three one topping pizzas 1 comes from the 1 in the "upper level" of the triangle, and two come from 2 in the "upper level".

That is why ya add 'em...

e-mail me soon with your response.

M I K E

Fig. 16.1 (*cont.*)

BACKGROUND

The students named in this article are part of a larger group that has thoughtfully and collaboratively investigated meaningful mathematical problems since grade 1. They have been invited to make conjectures and develop theories about their ideas. In the course of explaining solutions, they have justified their reasoning. We have had opportunities to reexamine students' original representations in the light of new discussions, as they revisited ideas in pursuing the same tasks or extensions of those tasks. We were particularly interested in examining how their ways of working, as shown, for example, by the representations they used and the *way* they used them, might change with time. We also wondered what connections they might make between earlier ideas and new problems.

In this article, we focus on the mathematical ideas of five students, Ankur, Brian, Jeff, Michael, and Romina, who have a long history of doing mathematics together. In elementary school, they were usually in the same mathematics class, if not the same group. They were observed within the classroom setting; at times, small-group follow-up interviews were also conducted. The high school mathematics sessions we describe took place after school, mainly on Friday afternoons. Tasks that they originally explored in the fourth and fifth grades initiated a series of related investigations involving combinatorics in the tenth and eleventh grades. Two problems initially posed by the research team in elementary school provided foundations for the explorations in the tenth and eleventh grades: The Pizza Problem and Building Towers. (The original pizza problem presented in grade 5 was the Pizza Problem with Halves: "A standard 'plain' pizza contains cheese. Toppings can be added to either half or the whole pie. How many choices are there if you can choose from two different toppings? List all possibilities. Convince us that you have accounted for all possibilities and that there could be no more." Students were then invited to consider this problem when selecting from four toppings.)

> *The Pizza Problem:* A pizza shop offers a basic cheese pizza with tomato sauce. A customer can then select from the following toppings: peppers, sausage, mushrooms, and pepperoni. How many choices for pizza does a customer have? List all the possible choices. Find a way to convince one another that you have accounted for all possibilities.
>
> *Building Towers:* Build all possible towers four (or five, or three, or n) cubes tall when two colors of Unifix cubes are available. Provide a convincing argument that all possible arrangements have been found.

At this point, the reader might want to pause a moment and review the mathematics that these tasks invite learners to consider. A pizza with four possible toppings, or a tower four blocks high (with two colors available), is determined by a *combination* of four things taken a certain number of times. For example, towers in two colors (say, red and blue) where exactly three of the four cubes are blue represent combinations of four objects taken three at a time, as do pizzas with exactly three toppings chosen from among four possible toppings. Both of these tasks challenge students not only to enumerate all possible combinations but also to *justify* their enumerations. The demand for justification, rather than simply for an "answer," sets the stage for building suitable representations through which lines of reasoning can be proposed and then explored, argued, and modified.

Archived videotape data, along with copies of students' work, gave us evidence of the students' mathematical ideas as fifth graders. At that time, they were part of a group of twelve classmates engaged in this problem investigation. Using a variety of strategies and representations, including a partial tree diagram, lists, and an organization that systematically controlled for variables, they were able to find all sixteen pizzas. Michael drew circles to represent the various pizzas, labeling each with its toppings. The students created codes using letters or abbreviations to represent the four toppings (e.g., *pep* for pepperoni, *m* for mushrooms), and they also decided to code for a pizza with no toppings (denoted, in different situations, either by *pl* for plain or by *c* for cheese). They distinguished between "whole" pizzas (plain or with one topping) and "mixed" pizzas (with two or more toppings). The original representations displayed by the students in grade 5 employed notation that enabled them to keep track of their ideas and to account for all possibilities to reach a solution.

Nearly five years later, the representations first displayed by Romina, Jeff, Brian, and Ankur were similar to those they had used in the fifth grade. They worked collaboratively, talking aloud about combinations of toppings and patterns that they were observing as they considered the cases of two, three, four, and five (modifying the original problem) available toppings. They initially built individual lists, using codes of letters to represent the different toppings, not unlike the codes they used in grade 5; however, they soon switched their notation to the numerals 1 through 5 so that they could more easily compare their lists.

Michael's Representation

Michael, however, spent at least fifteen minutes quietly developing his own solution, a symbolic representation based on a binary coding scheme. He joined the discussion when the other four students announced their solution. Michael disagreed with their assertion that if five toppings were available, thirty different pizzas could be made with at least one topping, plus one plain (or cheese) pizza, for a total of thirty-one.

> *Michael:* I think it's thirty-two, with that cheese [a plain cheese pizza]. And without the cheese, it would be thirty-one. I'll tell you why.
>
> *Ankur:* Mike, tell us the one we're missing then.

Michael responded by explaining what the zeros and ones meant in his representation and how they are used to write familiar numbers in base two.

> *Michael:* Okay, here's what I think. You know like a binary system we learned a while ago? Like with the ones and zeros—binary, right? The ones would mean a topping; zero means no topping. So if you had a four-topping pizza, you have four different places—in the binary system, you'd have—the first one would be just one. The second one would be that [*writes 10*]; that's the second number up. You remember what that was? This was like two, and this was three [*writes 11*].

Jeff recalled where they had seen this before.

Jeff: I know exactly what you're talking about. It's the thing we looked at in [eighth grade]; it was with computers.

Michael continued to relate his coding scheme to the pizza problem.

Michael: Well, you get, I think—I have a thing in my head. It works out in my head. You've got four toppings. This is like four places of the binary system. It all equals up to fifteen. That's the answer for four toppings.

When Romina requested clarification, Jeff was able to respond.

Romina: So is the *one*—is that your topping?

Jeff: Yeah. Each one is a topping. The zeros are no toppings. The ones are toppings.

Michael then summarized, with acknowledgment from Brian.

Michael: So you go from this number [*indicates 0001*], which is in the binary system, it's one, to this number [*indicates 1111*], which is fifteen, and that's how many toppings you have. There's fifteen different combinations of ones and zeros if you have four different places.

Brian: Wow!

Michael: I don't know how to explain it, but it works out. That's in my head—these weird things going on in my head. And if you have an extra topping, you just add an extra place and that would be sixteen, that would be thirty-one.

At this point, Michael's representation allowed him to account for all pizzas having at least one topping. Although the students later realized a code of all zeros represented a plain cheese pizza, this was treated as a separate case initially.

Jeff: And then you add the cheese?

Michael: Plus the cheese would be thirty-two.

With the assistance of other students, Michael presented his binary coding scheme to the teacher-researcher, saying, "This is the way I interpret it into the pizza problem."

Teacher-researcher: What's the difference between 1-0-0-0 and 0-1-0-0?

Jeff: Well, that would be the difference between an onion pizza and a pepperoni pizza.

Michael had written his codes in a vertical list. Jeff suggested labeling each column with the name of a topping. Michael agreed, noting that a 1 in a column indicated that that pizza had that particular topping. The students were then asked to consider the cases of ten toppings and, finally, n toppings. After much discussion, they concluded that the number of possibilities for n toppings must be 2^n.

The teacher-researcher then asked the students whether this problem reminded them of anything else that they had done before.

Brian: Everything we do always is like the tower problem.... Instead of building a tower, you're building a pizza.

Remembering their earlier explorations of block towers, begun in grades 3 and 4, the students discussed how towers might relate to pizzas.

Ankur: With the towers, you have like red on top and yellow on the bottom and then yellow on top and red on the bottom ... but with the pizzas you can't have peppers and pepperoni and then pepperoni and peppers. Understand that?

Teacher-researcher: So you're saying it's not like it?

Michael: It's similar.

Ankur: It's similar, but it's not exactly alike.

They compared the enumeration of pizzas with three available toppings to that of three-high block towers with two colors. This time, Romina, Ankur, Jeff, and Brian worked together to list the towers, using a code of "Y" for yellow and "R" for red, while Michael worked alone.

Jeff: How many you got over there, Mike?

Michael: I don't know. I'm just writing my binary again 'cause it might work for this!

Several minutes later, the students agreed that there would be eight towers of height three when they could select from two different colors, just as there were eight different pizzas when three toppings were available. They also noted sixteen towers of height four, again selecting from two colors, just as there were sixteen pizzas when four toppings were available. Intrigued but tired after working for more than two hours, they decided to adjourn.

One week later, all five students worked together to construct an argument that the pizza and towers problems are alike, using Michael's binary notation to construct a mapping between pizzas and towers.

Michael: Two colors? All right, that, that's easy, 'cause you can still use the binary ... say I made "zero" blue and "one" red, and it's the same thing—you can have any number of combinations.

Presenting at the board, Michael drew a chart with column headings "mush" and "peppers" and listed his binary representations for the four possible pizzas when two toppings are available. He then modified his chart, replacing "mush" with "2" and "peppers" with "1." The column headings then indicated the two positions that a block can occupy in a two-high tower, rather than pizza toppings.

Michael: You understand, if you have a tower with two colors, that would be the same as the pizza problem.

The students explained that the number of solutions for both problems, counting n-high towers or n-topping pizzas, is 2^n, where the symbol n represents either the number of pizza toppings or the height of the towers. The number 2 denotes, accordingly, either the presence or the absence of a given pizza topping or the number of colors (two in this example) available for each position in a tower of given height.

A full year later, in the task-based interview, we invited Michael to explain once again his binary representation with respect to towers three-high and pizzas with three toppings. We were interested in how he saw these problems now and how his binary approach might connect to further contexts.

Michael: Between the 000 and the 111, there's every possibility you could—you give me one, it's in there—with three. So, that's what I like about the binary system, that everything's in there. So you take [*writes 000 to 111*]—and each one is a number.

Michael was able to easily count the total number of possibilities for a situation because he saw these representations as numbers, always writing his lists sequentially, for example, from 0000 to 1111, often saying the base-ten equivalents out loud as he wrote. For an individual representation, such as 101, the meaning of each 1 and 0 depended on the context, such as a presence or an absence of a particular topping on a pizza or the location of a cube of a certain color in a tower.

In the concluding paragraphs of his e-mail, Michael explained why Pascal's triangle represents the pizza problem by demonstrating in a particular case that the addition rule that determines the entries of Pascal's triangle must also hold for pizzas. In this written exposition, he illustrated the general process informally, moving from pizzas with four toppings to pizzas with five. In a subsequent after-school session, however, when invited to explain his ideas to the group, Michael gave an argument for passing from pizzas with n toppings to pizzas with $n + 1$ toppings (Kiczek 2000). Whether presented informally or formally, Michael's argument is conceptual because it treats the numbers that appear, both in Pascal's triangle and in the pizza problem, as the results of underlying structures that he demonstrates are mathematically equivalent.

In the group discussion prior to Michael's interview and subsequent e-mail, the binary code was already linked explicitly to towers. Hence we may also read the interview and the e-mail as demonstrating, more generally, that block towers of different heights are also modeled by Pascal's triangle.

CONCLUSIONS

The towers and pizza tasks became metaphors for thinking about important mathematical ideas. Michael's representation enabled him to list all possibilities, first for pizzas and then for a special case of towers. On the basis of this systematic organization, Michael built arguments, first informally for special cases and later in greater generality. Michael worked alone before he shared his coding scheme with others. When he did share it, despite apologies ("I don't know how to explain it, but it works"), Michael showed quite clearly how he used the code to solve the given problem. Although the other students acknowledged its usefulness in later problem situations, the code continued to be called Michael's "binary thing."

Michael's representation resurfaced later, as a means of keeping track of different sequences of possibilities, given two choices at each stage, such as on/off, red/blue, heads/tails, and win/loss. Further, Michael and his peers were able to extend and modify the coding scheme substantially in order to solve problems such as counting towers when *three* colors are available (Muter 1999) and to model certain sample spaces for probability problems (Kiczek 2000).

Detailed study of the videotapes of the sessions described here reveals students who are accustomed to thinking about mathematics and to communicating ideas. They work comfortably with one another and are able to bring multiple perspectives to problem explorations, with very little intervention by a teacher. Throughout the investigations, they cycled through a complex, many-sided process of discussion and reflection, often "folding back" (Pirie and Kieren 1992) to earlier ideas. Through their individual and group analyses, they presented, discussed, and modified a variety of strategies and observations. Further, providing justification for their conclusions each step of the way helped the students become more confident about their findings. Years of shared mathematical problem-solving experiences nourished rich discussions, in which the students reconsidered and reshaped their earlier ideas, built new images and representations, and then tackled further problems, sometimes of their own design. Their mathematical power emerged in an environment that invited students to confront and reconsider prior images, interpretations, and hypotheses. Our work strongly suggests that appropriate investigations, extended and revisited over long periods of time, can help students build strong foundations for important thinking in which representations, images, and proofs seem to emerge *together,* in a single process.

We have reported the work of five students who together made some deep connections among important mathematical ideas. For these students, the properties of combinations grew from very concrete images, such as towers and pizzas. Once these properties emerged in Pascal's triangle, they linked those images into a larger framework that connected quite readily to ideas in algebra and probability. As such, the prior images anchored abstractions (Maher and Speiser 1997) that were built as students folded back to reconsider and revise earlier images while linking them to new ones that served different functions. Michael's representation, triggered by the need to find and justify a particular solution, served as a tool for him and others to connect mathematical situations that he and his classmates explored for a number of years.

Our work is rooted in the belief that students can build lasting mathematical images and understandings when they are presented with carefully designed tasks that invite them to investigate rich problem situations (Davis and Maher 1997; Maher and Speiser 1997). The investigation of particular tasks that elicit the construction of combinatorial and probabilistic ideas forms one important strand in the longitudinal study. Combinatorial investigations have been found useful for studying the growth of mathematical thinking (Maher, Muter, and Kiczek in press; Maher and Martino 1992, 1996; Speiser 1997). Such explorations invite students to build ideas and strategies that they can then test in new situations, which leads to cycles of conjectural extension, reconsideration, and reformulation.

IMPLICATIONS

The students worked in a community where ideas were shared, discussed, and revisited, often over long periods of time. They were invited to reconsider earlier problems in contexts that offered opportunities for rethinking and for further building. This study indicates that when certain conditions are in place, students can and do construct convincing arguments to support their findings. The search for justification triggers the construction of new images and representations, which, in turn, serve as anchors for further connections and abstractions. The development of mathematical reasoning cannot be the objective of a single activity but rather emerges in the course of long-term processes.

The tasks we have described are parts of specific strands of mathematical investigations that have been under development in the course of research for more than a decade. To build important mathematical understandings through long-term investigation, the tasks and situations that learners explore must be rich enough to encourage the learners to develop personal heuristics, techniques, and strategies and then share them with their peers, in conversation, classroom presentations, and writing. As teachers, we respond in ways that help the learners widen their repertoire, extend their thinking, and revisit earlier ideas. The act of writing promotes reflection, editing, and reorganization. Revisiting gives students the opportunity to reconsider other ideas, which are often worthwhile alternatives. Inventing a personal solution is insufficient; it is also necessary to contemplate other ways of thinking.

As a practical matter, most classroom teachers do not have the opportunity to work closely with the same students over the course of several years. However, it is possible for teachers within a school to collaboratively develop particular tasks and activities that students may explore and revisit at various grade levels, thereby offering experiences that will foster the development of mathematical understanding. The example of Michael's representation challenges us all to reflect on our current practice and to consider what might be possible in mathematics classrooms when suitable conditions are in place.

REFERENCES

Davis, Robert B., and Carolyn A. Maher. "How Students Think: The Role of Representations." In *Mathematical Reasoning: Analogies, Metaphors and Images,* edited by Lyn D. English, pp. 93–115. Mahwah, N.J.: Lawrence Erlbaum Associates, 1997.

Kiczek, Regina D. *Tracing the Development of Probabilistic Thinking: Profiles from a Longitudinal Study.* Ed.D. diss., Rutgers, The State University of New Jersey, 2000.

Maher, Carolyn A., and Amy M. Martino. "Teachers Building on Students' Thinking." *Arithmetic Teacher* 39 (March 1992): 32–37.

———. "Young Children Invent Methods of Proof: The Gang of Four." In *Theories of Mathematical Learning,* edited by Leslie P. Steffe, Pearla Nesher, Paul Cobb, Gerald A. Goldin, and Brian Greer, pp. 431–47. Mahwah, N.J.: Lawrence Erlbaum Associates, 1996.

Maher, Carolyn A., Ethel M. Muter, and Regina D. Kiczek. "The Development of Proof Making by Students." In *Theorems in School,* edited by Paolo Boero. Dordrecht, Netherlands: Kluwer Academic Press, in press.

Maher, Carolyn A., and Robert Speiser. "How Far Can You Go with Block Towers?" *Journal of Mathematical Behavior* 16 (1997): 125–32.

Muter, Ethel M. *The Development of Student Ideas in Combinatorics and Proof: A Six-Year Study.* Ed.D. diss., Rutgers, the State University of New Jersey, 1999.

Pirie, Susan E. B., and Thomas E. Kieren. "Watching Sandy's Understanding Grow." *Journal of Mathematical Behavior* 11 (1992): 243–57.

Speiser, Robert. "Block Towers and Binomials." *Journal of Mathematical Behavior* 16 (1997): 113–24.

Part 4: The Role of Context

17

Listen to Their Pictures
An Investigation of Children's Mathematical Drawings

Kristine Reed Woleck

REPRESENTATIONS are not static products; rather, they capture the process of constructing a mathematical concept or relationship. Representations allow mathematicians to record, reflect on, and later recall their process and thinking. They become tools to turn to as mathematicians articulate, clarify, justify, and communicate their reasoning to others. These representations may take a variety of forms, from diagrams and models to graphs and symbolic expressions. Despite this broad view of representation, virtually no research has focused on one tool of mathematical representation and communication that is particularly significant to the youngest mathematicians—the mathematical tool of drawing.

Drawing can be a window into the mind of a child. From a developmental perspective, drawing is a form of graphic symbolism that develops prior to writing (Dyson 1983). Vygotsky (1978) also refers to drawing as a "preliminary stage" in the child's development of written language. Drawing emerges as a powerful medium for discovering and expressing meaning; for the young child, drawing brings ideas to the surface. Add to this the notion that drawing itself involves the creation and manipulation of symbols, an essential component of logico-mathematical development, and the significance of children's pictures becomes all the more evident. Yet, in the realm of early childhood mathematical representation and communication, the child's drawing of pictures as a "tool" for understanding mathematics and as a "language" for sharing mathematical work has received little attention.

In my own first-grade, public school classroom in southwestern Connecticut, I undertook an action-research study to contribute to this limited body of research. I investigated the question, How do first graders make use of pictures as they represent and communicate their mathematical understandings? In the pages that follow, I use children's drawings and classroom anecdotes to describe and document this investigation. After a peek into the classroom context that set the foundation for the study, I discuss the children's functional and dramatic uses of drawing. I then show how the drawing process fostered the development of abstraction and symbol use for this group of first graders. Before closing, I let us hear the talk that surrounded the children's pictures; this talk can be the most compelling and valuable aspect of mathematical drawings for both student and teacher. When we open our eyes to the drawings of children, we open our ears to the richness of their mathematical voices.

SETTING THE STAGE

I began my research by setting the stage for the children's mathematical representations through meaningful mathematical experiences, investigations, and problem-solving situations. If children are to represent, record, and communicate mathematics, then they must have rich mathematical experiences to explore, ponder, and share. The young mathematicians of Room 1R (as we called our classroom) were given time to explore manipulatives and solve open-ended problems that encouraged flexibility of thought and a sharing of diverse strategies and solutions.

We also set the stage in the classroom by talking about mathematics. Drawing and all other sign systems are built on a foundation of verbal speech (Vygotsky 1978); oral language precedes and is a bridge to graphic representation (Dyson 1983). Thus, we talked about the different strategies that members of the classroom used to solve a particular problem. We talked about the mathematics happening around us, in the books we read and in the routines of our daily life. We talked about the mathematics happening in photos and pictures found in newspapers and children's literature. These discussions of the mathematics embedded in pictures and illustrations explicitly validated drawing as a form of representation and communication used by mathematicians. With this stage set, Room 1R mathematicians certainly had plenty to represent, record, and share.

PICTURES WITH A PURPOSE

It became evident early on in my research that pictures could be used by children in distinctly different ways to support mathematical learning and communication. At times in the classroom, pictures were used in a functional manner to support problem-solving efforts. Some of the first graders used their pictures as if they were manipulatives to be organized and counted to solve a problem. Like concrete manipulatives, these pictorial representations served as placeholders for thoughts, which allowed the children to carry out and record the intermediate steps of a task without being overwhelmed by the rest of the problem (NCTM 2000).

This functional use of pictures was evident in Dominique's work to solve "My Father's Wheel Problem" (fig. 17.1). In this open-ended problem, children brainstormed the various means of transportation that three children could take to school and then determined the total number of wheels used. When solving the problem, Dominique drew the three children and the wheels of the roller skates that he had chosen as their means of transportation. The wheel drawings became a mathematical tool, used like Unifix cubes or other counters; Dominique counted the wheels on his paper one by one to arrive at his solution of 14. When sharing and "proving" his work to others in the group, he pointed to each wheel he had drawn as he counted by ones. In this way, pictures can serve as a tool—a scaffold erected by and for the learner—to support problem-solving efforts and the development of

Fig. 17.1. Dominque's drawing for "My Father's Wheel Problem"

mathematical concepts. The pictures can serve as representations for the purpose of learning and doing mathematics (NCTM 2000, p. 67).

At other times, however, pictures can serve not as a functional representation, but rather as a "dramatic" representation to accompany mathematical thinking and to communicate the essence of the mathematical work. Ryan produced such a dramatic picture in mathematics as he recorded his solution strategy for the problem, How many cups will we need for the Author's Breakfast? (fig. 17.2). To solve the problem by using the numerical data about the guests collected earlier that day in the classroom, Ryan pointed with a ruler to the relevant numbers on the classroom number line and counted on. His picture captured the physical drama of this work; it showed him pointing to the number line with the ruler. After drawing this dramatic representation, Ryan was able to organize his thoughts and articulate his

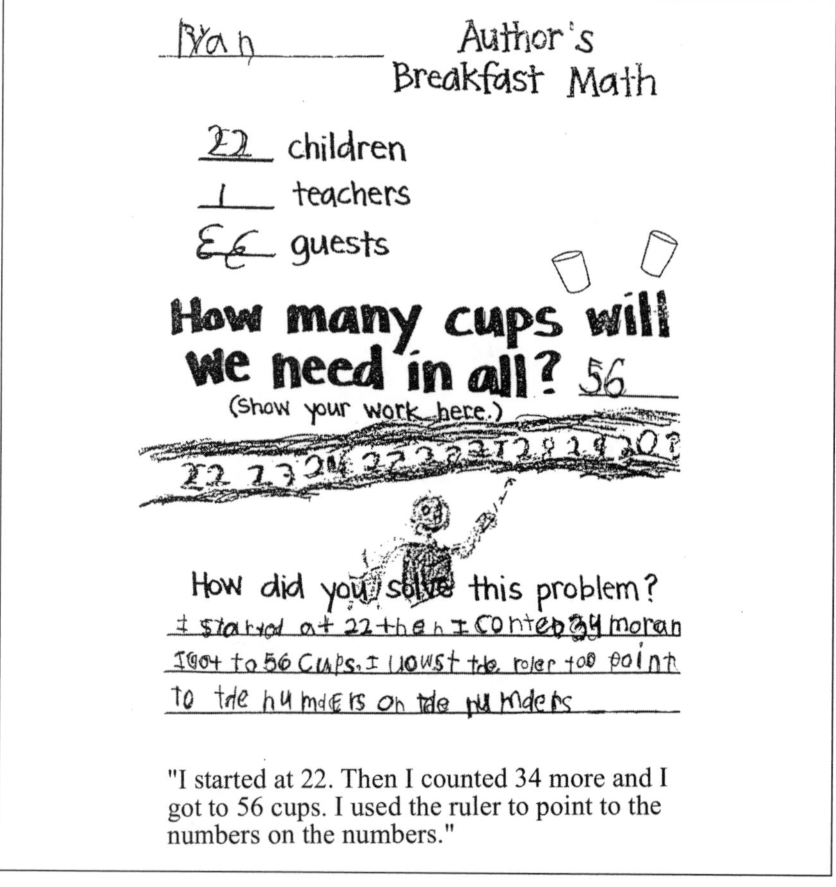

Fig. 17.2. Ryan's solution to the "Author's Breakfast" problem

strategy orally and in written words: "I started at 22. Then I counted 34 more and I got to 56 cups. I used the ruler to point to the numbers on the number line." Instead of serving as a functional problem-solving tool or as a "pictorial manipulative," Ryan's picture was a tool to support his mathematical communication—another purpose of pictures in mathematics. His dramatic representation demonstrates how a picture can serve as a prewriting tool that enables a young mathematician to reflect on and recall a mathematical process in order to communicate it to others.

Artist Meets Mathematician

As I observed the children attempting to represent their problem-solving efforts and mathematical thinking in the classroom, I witnessed how the children's use of pictures can be a forerunner to their use of symbols in mathematics. The symbols that children use in mathematics may not look conventional, but they are purposeful, are intentional, and carry meaning (Mills, O'Keefe, and Whitin 1996). In this way, as children invent and use their own symbols in their mathematical representations, they experience the sense-making quality that should underlie the use of all symbols in mathematics.

For William and Whitney, the "Witches' Transportation" problem prompted a meaningful transition from pictures to symbols. The problem (*Exemplars,* 1995–1996) presented the children with a complex situation to resolve:

> Twenty witches must travel on 8 brooms to a convention in California. They will have to "broom pool." No broom may carry more than 4 witches. No broom may carry fewer than 2 witches.
>
> How can they do it?

William immediately began drawing broomsticks that held witches with full bodies, faces, and hats. He became visibly frustrated, however, when he found that given the complex constraints of the problem, he repeatedly needed to revise his solution by erasing some witches and drawing more witches on other brooms. William then made a cognitive leap from pictures to symbols by drawing lines to represent the witches on each broom (fig. 17.3). He stated that he did not need to see the witches' bodies and hats and that it was "taking too much time" for him to draw them with such detail. Likewise, some of Whitney's first witches were drawn in full detail (two with facial features), but then she drew others with only a straight line (fig. 17.4). Whitney noted this drawing shift in her written words: "Then I made lines. It was a lot easier." Both of these children demonstrated a significant piece of cognitive development here; they came to separate a symbol from what it signified

(Piaget 1969). They became less bound to realism in their pictures of the witches and brooms and instead created their own symbols for these objects.

Fig. 17.3. William's solution to the "Witches' Transportation" problem

The work of William and Whitney documents a move from elaborate drawings to more symbolic representations. As these children moved from picture to symbol use, they engaged in abstraction. Abstraction in mathe-

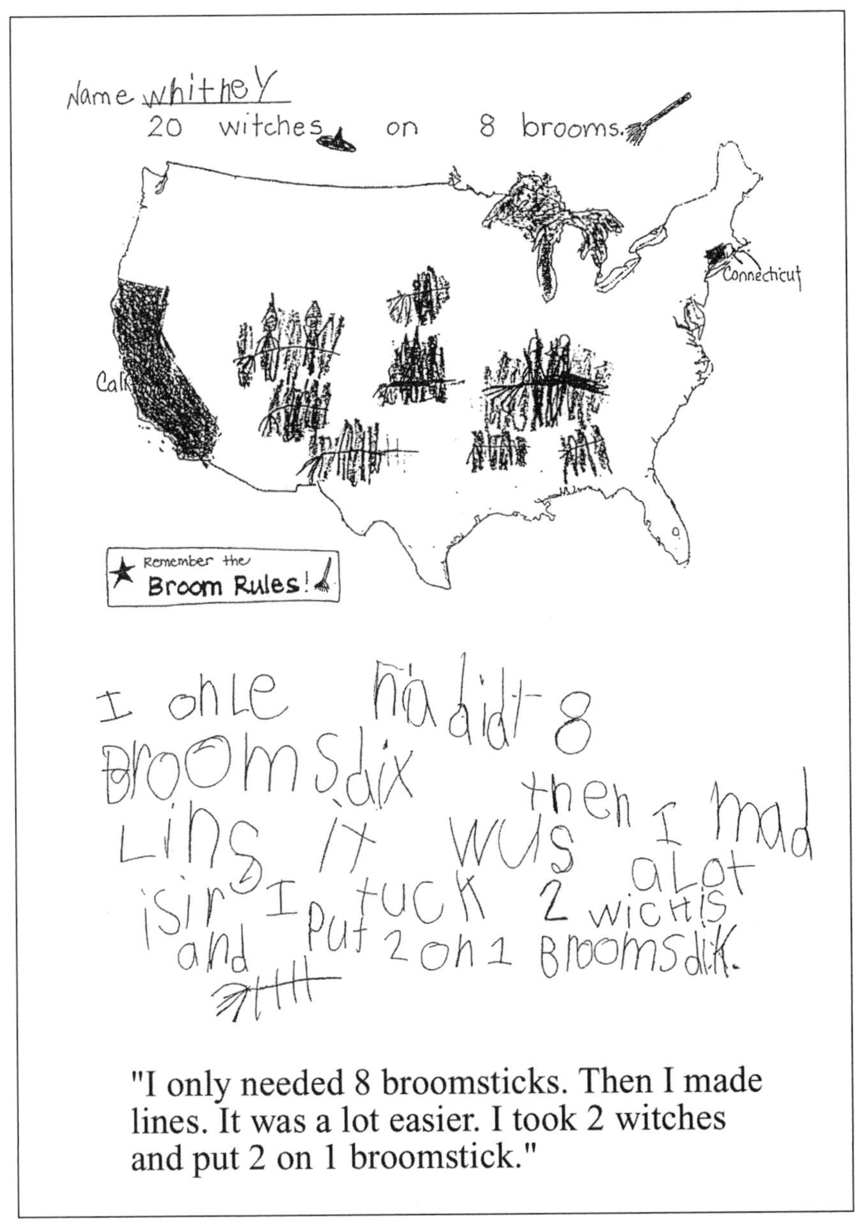

Fig. 17.4. Whitney's solution to the "Witches' Transportation" problem

matics entails using symbolization to strip away unnecessary features so that the symbols can be manipulated with greater ease (NCTM 2000, p. 69). William and Whitney stripped away the unnecessary details of faces, hats, and the like to leave the more symbolic, abstract lines that they could draw, erase, and cross out readily as revisions were needed. These children uncovered for themselves the important distinction between artistic drawing and drawing that serves as a mathematical tool.

SPEAKING OF PICTURES

As rich as these first-grade pictures and drawings may themselves appear, it was the talk surrounding the pictures that truly brought to life the mathematics embedded within each representation. "Self talk" could often be heard as children talked themselves through their process of drawing. For some, it was as if they were "living" the drama of their drawing, and it was a drama that could reveal much about a child's number sense to a teacher with an open ear. The following anecdote of Tim's drawing process and self-talk offers one example:

> After a whole-group reading of Mitsumasa Anno's *Anno's Counting Book* (1977), Tim chose to draw a page to illustrate the number seven (fig. 17.5). After several minutes of work, Tim pointed with a finger to eight purposeful marks on his paper and announced to himself, "Oh,

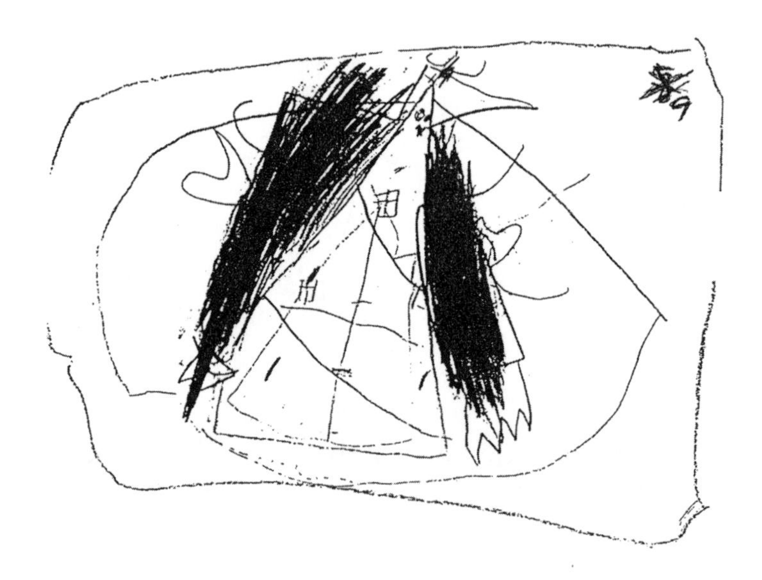

Fig. 17.5. Tim's illustration for the number 7, with revisions

one too much." He then crossed out the 7 at the corner of his paper and wrote the numeral 8 in its place. He began naming various marks on his paper (depicting a spaceship on a mission), "OK, so two engineers, five army generals, this guy's the pilot.... I need to change to 9. I need another pilot." With this, Tim crossed out the 8, wrote the numeral 9, and drew another mark, presumably representing his second pilot.

Tim's drawing is at an early stage from a developmental perspective. His work is affected by significant fine-motor issues. Thus, for teachers of children such as Tim, giving attention to the process of drawing is crucial. The teacher's open ear to the self-talk that guides the child's drawing is especially necessary in recognizing the mathematical thinking that the drawing process evokes for the child but that may not be readily apparent in the visual representation on the paper.

Talk also occurred among peers as they created mathematical drawings:

Kevin had drawn seven gifts as part of a problem-solving effort and was attempting to confirm that seven was an odd number (fig. 17.6). He pointed to each gift and spoke the pattern, "Even, odd, even, odd ...,"

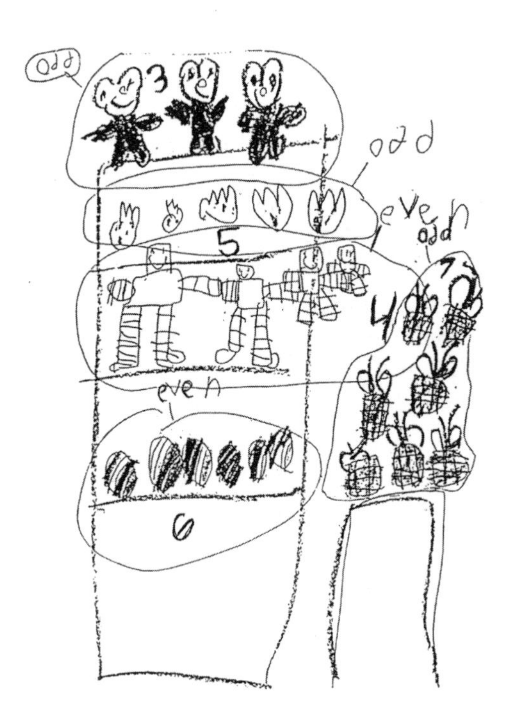

Fig. 17.6. Kevin's pattern of odds and evens

> beginning with "even" at the first gift. When he ended with "even" on the seventh gift, he paused with a look of confusion and turned to Parke (sitting next to him) for assistance. Parke looked at Kevin's drawings and counted pairs, linking two gifts each time with his fingers; he announced confidently, "It's odd 'cause there's a leftover." But Kevin was not yet satisfied and wanted to know why his "even, odd" pattern did not work. At that moment, Ryan, who had looked up during Parke's counting and overheard the conversation, said, "Hey, you can't start 'even, odd.' One is odd!" Kevin smiled and said, "Oh, right! It's odd, even, odd, even!" He counted his pictures again, this time with his new pattern, and proudly announced that seven was an odd number.

Such peer conversations were informal and spontaneous, but through talking together as they drew, these first graders came to question, debate, defend, clarify, and refine their mathematical understandings.

Often, these conversations among peers emerged during whole-group discussions of their drawings. During these discussions, children shared their work, while I posed careful questions, when appropriate, to foster mathematical thinking and communication:

> David brought to the whole-group circle his illustration of the number 7. It included a set of seven flowers growing on the ground and an airplane decorated with seven stars. He attempted to count the seven stars by twos. He stopped at 6, puzzled by the remaining one. Again, he counted by twos, this time stopping at 8 with a look of confusion. I asked the group, "What's happening here? David is counting by twos and keeps getting to 6 and 8. Why can't he get to 7?" An animated discussion erupted as various mathematicians in Room 1R went to David's picture to explain their thoughts on this matter. Ella attempted to articulate the notion of skip-counting as she pointed to the set of items in David's picture, "because when you count by twos you skip 7, like you skip 1 and say 2." Jay interjected, "And skip 3, 5, 7." Julie continued, "You count 2, 4, 6. Just count one more then! 7!"

David's picture thus served as the stimulus for some constructive mathematical talk that also gave me valuable insights and assessment data about the children's thinking. I could then use this information to shape my teaching.

These classroom examples show clearly not only that a picture alone may indicate something of a child's mathematical understanding and communication, but also that the richness of the child's thinking is lost if the picture is considered out of context or is alienated from the "who" of the mathematician. Pictures are not simply static end products to be collected and examined; they are dynamic tools to support the child's mathematical thinking and communication with others in the classroom community. It is the language the child attaches to these pictures that gives teachers insights into the child's understanding and thinking. It is the talk that these pictures stimulate that prompts young mathematicians to revisit, challenge, and clarify their ideas. The pictures serve as springboards for the talking of mathematics, which in turn brings about still more learning and doing of mathematics for the young child.

DRAWING IT ALL TOGETHER

Conscious of the focus of my research, I attempted at times to structure mathematical tasks, problems, or questions in a manner that I believed would prompt or encourage the use of pictures as a tool of representation and communication. Interestingly, though, I found over time that it was more valuable to simply examine the use of pictures in the full context of the children's ongoing, daily work in the classroom. Rather than attempt to isolate or emphasize drawing artificially, I came more and more during the fall to note how pictures naturally functioned with the words, numbers, and other mark-making activity of the children's mathematical work. I found that children move among these many tools in an integrated, fluid process to represent and communicate mathematics. It is as if children themselves come to make use of drawings in situations that render them necessary; it is as if these young mathematicians often have an intuitive sense of the language their pictures can "speak" in mathematics. Samples of Jessica's work (fig. 17.7) and Kevin's work (fig. 17.8) illustrate this integrated and fluid use of pictures. For the young child, pictures do not function alone, but rather "make their mark" in conjunction with words, numbers, and other tools of mathematical representation and communication.

This research thus comes full circle; it has specifically examined the pictorial work of the young child through the lens of mathematics, but it has also pulled back to place these pictures in their broader context with other modes of mathematical representation and communication. The wealth of insights gained from the examination of these pictures points to the need to explicitly validate the use of pictures in the early childhood classroom. These pictures should not be treated as static, extraneous, or second-rate pieces of mathematics work. They should be extolled as valuable tools for the learning and doing of mathematics and should be employed as springboards for the sharing and talking of mathematics. Pictures can then prove to be essential to both student learning and teacher assessment, as so many of the classroom examples here have shown.

At the same time, drawing must be permitted to function in the total context of a child's mathematical work. Mathematics is a living language, one that gains its power from the integration of oral, pictorial, and written

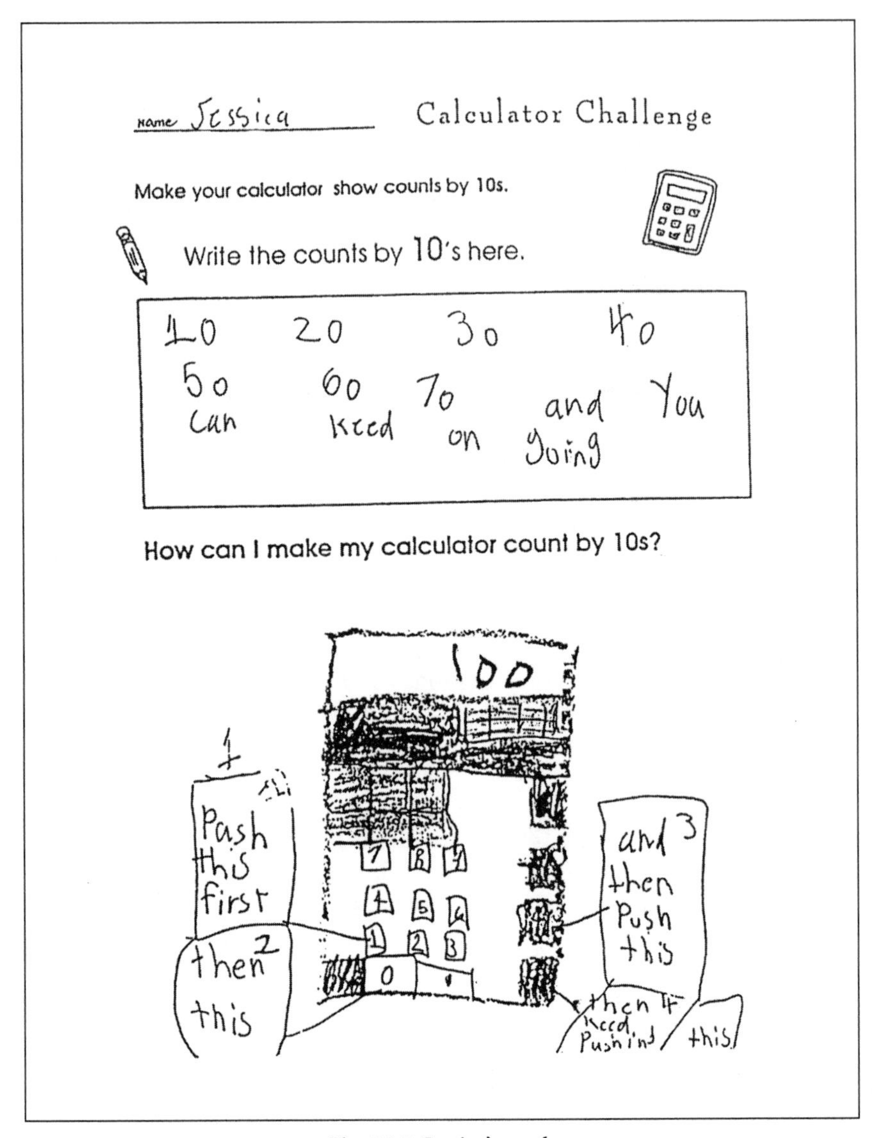

Fig. 17.7. Jessica's work

modes. Each form of representation and communication has its own strengths. A picture may best serve one purpose; words or symbolic expressions may best serve another. Children need opportunities to grow fluent in the integration of all these forms of symbolism, representation, and communication. Only then can the voices of young mathematicians truly be heard.

Fig. 17.8. Kevin's work

REFERENCES

Anno, Mitsumasa. *Anno's Counting Book.* New York: Harper-Collins, 1977.

Dyson, Anne Haas. "Research Currents: Young Children as Composers." *Language Arts* 60 (October 1983): 884–89.

Mills, Heidi, Timothy O'Keefe, and David Whitin. *Mathematics in the Making: Authoring Ideas in Primary Classrooms.* Portsmouth, N.H.: Heinemann, 1996.

National Council of Teachers of Mathematics (NCTM). *Principles and Standards for School Mathematics.* Reston, Va.: NCTM, 2000.

Piaget, Jean. *The Child's Conception of Physical Causality.* Translated by Marjorie Gabain. Totowa, N.J.: Littlefield, Adams, & Co., 1969.

Vygotsky, L. S. *Mind in Society: The Development of Higher Psychological Processes.* Cambridge, Mass.: Harvard University Press, 1978.

"Witches' Transportation," *Exemplars* 3, no. 2 (1995–6): 1–8.

18

Using Literature to Invite Mathematical Representations

Phyllis Whitin

David Whitin

Encouraging children to generate their own visual representations of mathematical patterns and relationships has many benefits (Atkinson 1992; Folkson 1996; Groves and Stacey 1998; Whitin 1997). Creating such visual counterparts is an act of sense-making that challenges children to organize and display significant mathematical features of an experience. When given the opportunity to display mathematical ideas in their own way, children display remarkable insight and inventiveness. One context that supports these student-generated representations is the realm of children's literature. Stories give mathematics authentic contexts and offer the potential for discussion, exploration, and extension (Whitin and Wilde 1992, 1995). In fact, open-ended conversations about stories set the stage for children to create representations that are diverse in nature. This article describes one such experience with a class of kindergarten children who used a counting story to investigate mathematical patterns.

Reading and Talking about the Story

In the spring we involved the children in several experiences with skip-counting. We used stories and rhymes to highlight counting by twos: "One, Two, Buckle My Shoe," *One, Two, One Pair!* (McMillan 1991), *Who Took the Cookies from the Cookie Jar?* (Williams 1995), and *How Many Feet in the Bed?* (Hamm 1991). We invited children to create their own visual representations in response to the mathematical pattern in this last book. The story invites readers (and listeners) to count the feet of five family members as they tumble in and out of bed on a Sunday morning. Dad, alone in bed, begins the count with "two," and he is joined successively by his daughter, son, baby, and wife. However, a series of incidents, such as the ringing of the phone and the overflowing of the bathtub, force family members to leave the bed one by one—or two by two, if you're counting feet!

An interesting conversation occurred during the reading of the story. Since picture books themselves are visual representations, children's oral responses reflect their observations of this visual display, their understanding of basic story structures, and, in this instance, their knowledge of the mathematical pattern represented. This conversation was also important because it was one of the factors that influenced the representations that the children later created.

As we read aloud about the first two people in the bed, the children chimed in: "I see four feet," "We have two feet," and "It's going by two." We then encouraged the children to predict what would happen as more family members joined the group. When the number of feet equaled ten, we held up the book and said, "Look, the story is only half over. What do you suppose is going to happen in the next part?" This question asked the children to draw on their knowledge of how stories work—and mathematical pattern is one way that stories are put together. Jordan responded, "It's going to go backwards: 10, 9, 8, 7, 6, 5, 4, 3, 2, 1." Brittaney added, "It went Dad, then daughter, little boy, baby, and Mom. Mom is going to be gone first." She built on Jordan's idea of a descending order and added the specific sequence of family members. When we asked her to "say some more about this idea," she added, "They'll be taking turns, taking turns going first." Her remarks underscored the symmetry of the comings and goings in this story. She highlighted the literary structure of the story in a mathematical way.

After several other children offered predictions, Savannah returned to Jordan's numerical idea. She suggested, pausing between numbers, "It would go 9 ... 7 ... 5 ... 3 ... 1." We asked her, "How did you know it would count this way?" She reasoned, "10, 9, 8, 7, 6, 5, 4, 3, 2, 1. It won't be the same"—that is, as counting by ones, as Jordan had proposed. "Jordan said, '10, 9, 8.' I know that's not right because it goes by 2s." Thus, Savannah challenged the ideas of another classmate but in a respectful way. This is part of the culture of her mathematical community: to question ideas and to use logic and reason as the currency for conversation (Hiebert et al. 1998).

This portion of the conversation highlighted some important aspects of the teacher's role. The teacher should not pass judgment on the students' predictions but invite them to share their thinking. What teachers choose *not* to say is just as important as what they choose to say. We did our best to encourage these multiple predictions, knowing that the story would give evidence for or against them as the reading continued. It was not necessary to correct the students' predictions. In fact, as we began to read the rest of the story, Savannah revised her thinking entirely on her own. When we read that eight feet were left in the bed, she said aloud, "That will be 9, 10," specifying the two numbers now gone. As the story progressed, she said, "8 and 7" to show that six feet remained. Then she predicted the rest of the pattern: "Next will be 5, 6," and "Now there are 3, 4. Then there will be 2." The unfolding story supported her in this revision of her number pattern.

This conversation illustrates the particular classroom culture that produced the visual representations described in the next part of the article. This culture supported predicting, searching for patterns, and making personal connections. It also fostered certain dispositions about learning. Students discovered the importance of taking risks, revising one's thinking, and extending and challenging one another's ideas. These values and dispositions are the unwritten rules and hidden customs of classroom life that set the parameters for learning possibilities. They restrict or expand the ideas that children are willing to consider. Consequently, we cannot separate visual representations from the talk that surrounds them or the larger cultural context in which they are conceived. The exploratory nature of this conversation supported the diversity of the students' visual responses, and the depth and range of their resulting mathematical ideas.

CREATING VISUALS IN RESPONSE TO THE STORY

In order for the children to make sense of the counting-by-twos pattern for themselves, we wanted each of them to create his or her own visual response to the story. We suspected that the pictures in the book, the story line, the children's personal experiences, and our conversation would all influence their drawing and writing. After reading the story aloud, we asked the children to name other body parts that came in twos. They named eyes, ears, arms, hands, and legs. We then invited groups of three, four, and five children to stand, and we counted the different sets of two. This acting out of the story served as a personal connection to the visual and numerical information in the book. One of us then moved to the easel and said, "In a few minutes everyone will get a piece of paper to write and draw about the twos pattern. I'll do mine first. I think I'll make mine about four of my children's cousins." He proceeded to draw the four children, writing their names below their pictures. "How many eyes do these cousins have?" he asked.

Several children chorused, "eight," but we were intrigued by Savannah's response. Instead of counting by twos, she said, "Four plus four." At first we thought that she had grouped the four children in sets of two, but when we asked her to explain her response, she pointed to one eye on each child and counted: "One, two, three, four." Then she went through the drawings again, counting the other eyes, "Five, six, seven, eight." Later we saw her interest in this "twice around" relationship reflected in her own drawing.

With the children's help, we then wrote a story using the pattern of twos represented on the easel. "I have four cousins," our story began. "They have

8 eyes. 2 + 2 + 2 + 2 = 8." To emphasize the pattern, we wrote 2, 4, 6, 8, putting one numeral under each cousin.

Next we invited the children to create their own story of twos using drawing, writing, and numbers. We purposely gave the children blank paper for their work. If we had specified one area for drawing figures, another for recording numbers, and a third for writing, their products probably would have looked very much the same. We thought that a diversity of responses would help us assess the children's thinking, and we wanted to provide the children with an opportunity to make sense of the numerical sequence in their own way. Although many children used portions of our demonstration at the easel, they also created unique forms of representation as they wrote and drew.

Trey focused on a different aspect of the problem (fig. 18.1). After writing out 2, 4, 6, 8, 10, he carefully circled each numeral. When we asked him to tell us about the circles, he replied, "I made circles so the numbers won't get mixed up." His response showed that he was considering the audience for his mathematical text. He was concerned that if he did not clarify his meaning by separating the numbers in the sequence, his intentions might be misinterpreted by others. Although he did not know the mathematical convention of using commas to separate numerals, he did have a clear understanding of its purpose. Just as children will invent their own ways to add punctuation in writing (Harste, Woodward, and Burke1984), Trey was inventing a convention to help him communicate mathematical ideas.

Kinley's picture highlighted another mathematical aspect of the story (fig. 18.2). She decided to keep track of both the number of eyes (6) *and* the number of people (1 + 1 + 1 = 3). When children are given open-ended

Fig. 18.1

Fig. 18.2

opportunities, they will pick up on information that is interesting to them, creating new possibilities by extension. In retrospect, we wish we had capitalized on Kinley's response to explore the following pattern:

1 person	2 eyes
2 people	4 eyes
3 people	6 eyes
4 people	8 eyes
⋮	⋮

Inviting Kinley and her classmates to consider the numerical relationship in this new form might have led to some exciting discoveries.

Other children responded to the experience's open-ended nature in different ways. Some added other kinds of numerical information. Amanda noted, "My cousin is 8," and Michael wrote, "My mom has 11 cousins," although he chose to draw members of his immediate family for the twos sequence. Tyler and Rachael extended their stories by supplying feelings of affection in statements like "You are my bestest friends" and "My babysitters are nice to me." Their work demonstrates that mathematical thinking operates in the context of real-life experiences.

Two of the children's visual representations corresponded directly with the conversation during the read-aloud session. Savannah drew five of her cousins (fig. 18.3) and placed the total number (10) on the top of her paper along with information about the eyes and the people. We asked her to say more about her work. She counted one eye on each person, and then returned to the first figure and began counting second eyes by ones again to reach 10. Her visual representation reflected the strategy that she had used during the acting-out portion of the lesson. Savannah had shown an understanding of skip-counting forward and backward during the story reading, but now she was apparently interested in showing the numerical relationship in a different way. In order to capitalize on this interest, we asked, "How could you use numbers to show how you counted?" Savannah returned to her drawing and placed one numeral (1, 2, 3, 4, 5) by each figure, and then she continued with the second half of the numerals (6, 7, 8, 9, 10). Just as with Kinley, we subsequently wished that we had pursued Savannah's idea. If we had displayed her work in chart form, as shown here, Savannah and her classmates may have found additional interesting relationships.

1	2	3	4	5
6	7	8	9	10

Nevertheless, we believe that Savannah benefited from reconstructing the "twice around" strategy in written form.

Brittaney also connected the class discussion to her visual representation. At the halfway point in the story that we read aloud, she had predicted that the family members would leave the bed and that the numbers would then go backwards. Now she drew a family of four and created her own story (fig. 18.4). Using invented spellings, she wrote beside the first family member, "If you pot whan [put one] more it will be 4 [eyes]." She repeated this process for each additional family member, increasing the number by two each time, until the point in her story at which one family member and then another goes to the grocery store. She then drew lines and arrows to show each

Fig. 18.3

Fig. 18.4

family member going away. Thus, she created her own visual symbols to capture this aspect of the story. When she showed her work to us, we asked, "How could you use numbers to show how the people left?" Brittaney thought and then wrote the numbers 1 through 8 arranged as shown here:

1	2	3	4	5	6	7	8
8	7	6	5	4	3	2	1

By placing the numbers in two distinct rows, she created an additional visual pattern to consider. "What do you notice now?" we then asked her. "They go forwards and backwards," she answered. We remarked that the numbers switch places—for example, 1 appears at the far left on the top row and on the bottom right of the second row, and so on. We believe that the story, the illustrations, and the class discussion piqued Brittaney's interest in skip-counting. Creating her own visual representation and narrative afforded Brittaney the opportunity to reconstruct the counting-backward pattern for herself. Like Trey, Brittaney also invented symbols (arrows and lines) to show a mathematical idea (decreasing sequence). Once she had her ideas recorded in visual form on paper, she was able to extend her thinking by talking with us. Visual representations, then, serve as placeholders for meaning as learners generate new ideas.

We were intrigued by both Brittaney's and Savannah's visual efforts. Both girls had participated actively in the discussion of the book, and both had created visual representations that were quite different from our demonstration. We began to wonder about the relationships among talking, drawing, and understanding. We now suspect that the more richly diverse the talk, the more richly diverse the representations will be, and we believe that those representations (and subsequent talk) will lead to enriched thinking. It is important for children to entertain a wide range of observations and ideas. Each child can then take the features of an experience that are most salient to him or her and reconstruct them through drawing and writing. We now wish that we had made a conscious effort to involve additional children actively in the conversation, and in further work with children we intend to pursue this hypothesis that richer talk leads to richer representations, which in turn lead to richer understanding.

As a postscript, we note that the children's fascination with counting by twos did not end here. We helped the children extend this initial experience in several ways over the next few weeks. One day, we led the children in a finger play called "Ten Little Candles on a Birthday Cake." The children held up both hands, showing all ten fingers. While reciting the rhyme, they successively "blew out" two "candles" at a time, represented by one finger on one hand and its counterpart on the other, reducing the candles from 10 to 8, to 6, to 4, to 2, and finally to none. Although the children had recited the rhyme periodically since the beginning of the year, no one had yet commented on the numerical progression. On the line, "And then there were six," Savannah's face lit up. "It's twos again!" she exclaimed.

On another occasion, the class constructed a Unifix-cube graph about favorite characters from *Who Took the Cookies from the Cookie Jar?* (Williams 1995), the centerpiece of another read-aloud session. The book features a rhyming story about a snake who eats cookies in pairs. Kinley used stacks of two cubes to count to ten. She opened up another mathematical opportunity by commenting that adding another cube would make the total an odd number. Another child, Kayla, returned to the book, and this time she noticed that the snake had two round humps in his body at the beginning of the story. She reasoned that the snake must have eaten two cookies before the story began, and therefore he had eaten twelve by the end of the story. What caused Kayla to make this new observation? We do not know for sure. However, we would argue that the cumulative experiences of looking at and discussing the visual patterns in stories and then creating one's own visual representations encourages readers to look more closely at the visual representations of other stories. Thus, in these ensuing weeks, the children represented twos using Unifix cubes, a finger play, and other pieces of literature. They recognized connections across stories, made distinctions between odd and even numbers, and interpreted a familiar story in a new way.

CONCLUSION

We believe that student-generated representations enrich mathematical understanding. Stories can often serve as catalysts for these representations. When children are given choice in how they represent their understanding, they create a diversity of representations. The children in our classroom had the opportunity to transform the ideas communicated by a story into their own personal representations. Transforming an idea involves rethinking, re-creating, and reconstructing it in a new form. Figure 18.5 represents the potential of stories and conversations to generate insightful representations and new understandings. We have seen, for instance, how Trey invented his own symbolic representation by using circles to separate his numbers. Children develop additional appreciation for traditional mathematical symbols when they first have the opportunity to create some themselves. The principle of commutativity loomed when some children added five sets of two to equal ten eyes, and Savannah added two sets of five (by going around twice) to get the same result. Kinley highlighted the numerical pattern of the number of people in relation to the number of eyes. Brittaney described another pattern by recording the reverse order of the characters in her story. Although we regret that we did not pursue some of these patterns with the rest of the class, the evident power of student-generated representations challenges us to capitalize on this potential in the future. Reflective teachers are constantly reviewing an unending series of cherished moments, missed opportunities, and promising plans for the future.

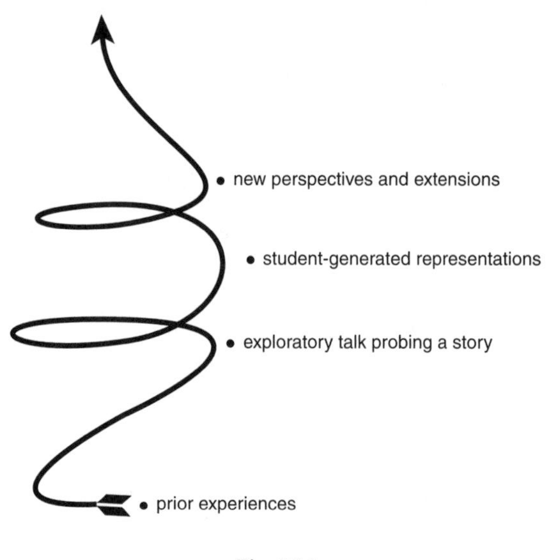

Fig. 18.5

REFERENCES

Atkinson, Sue. *Mathematics with Reason.* Portsmouth, N.H.: Heinemann, 1992.

Folkson, Susan. "Meaningful Communication among Children: Data Collection." In *Communication in Mathematics: K–12 and Beyond,* 1996 Yearbook of the National Council of Teachers of Mathematics (NCTM), edited by Portia C. Elliott, pp. 29–34. Reston, Va.: NCTM, 1996.

Groves, Susie, and Kaye Stacey. "Calculators in Primary Mathematics: Exploring Number before Teaching Algorithms." In *The Teaching and Learning of Algorithms in School Mathematics,* 1998 Yearbook of the National Council of Teachers of Mathematics (NCTM), edited by Lorna Morrow, pp. 120–29. Reston, Va.: NCTM, 1998.

Hamm, Diane. *How Many Feet in the Bed?* New York: Simon and Schuster, 1991.

Harste, Jerome, Virginia Woodward, and Carolyn Burke. *Language Stories and Literacy Lessons.* Portsmouth, N.H.: Heinemann, 1984.

Hiebert, James, Thomas Carpenter, Elizabeth Fennema, Karen Fuson, Diana Wearne, Hanlie Murray, Alwyn Olivier, and Piet Human. *Making Sense.* Portsmouth, N.H.: Heinemann, 1998.

McMillan, Bruce. *One, Two, One Pair!* New York: Scholastic, 1991.

Whitin, David. "Collecting Data with Young Children." *Young Children* 52 (January 1997): 28–32.

Whitin, David, and Sandra Wilde. *Read Any Good Math Lately?* Portsmouth, N.H.: Heinemann, 1992.

———. *It's the Story That Counts.* Portsmouth, N.H.: Heinemann, 1995.

Williams, Rozanne. *Who Took the Cookies from the Cookie Jar?* Cypress, Calif.: Creative Teaching Press, 1995.

19

Representation in Realistic Mathematics Education

Margaret R. Meyer

THIS article will explore a theory of mathematics learning and instruction and show how representation in middle school algebra might develop in an instructional sequence based on this theory. This theory, which is called Realistic Mathematics Education (RME), has evolved over the past thirty years of developmental research. RME represents a significant departure from traditional ideas about learning and teaching mathematics that can best be seen by examples that illustrate its principles.

PRINCIPLES OF REALISTIC MATHEMATICS EDUCATION

RME is based on five related principles of learning and instruction. The formulation of these principles here represents a synthesis of statements that appear in several sources (de Lange 1987, 1992; Treffers 1991). The five principles may be stated as follows:

1. Learning mathematics is a constructive activity, in contrast to one in which the student absorbs knowledge that has been presented or transmitted. Such construction becomes possible when the starting point of instructional sequences is experientially real to students, allowing them to engage immediately in mathematical activity that is personally meaningful.

2. The learning of a concept or skill is accomplished over a long period of time and moves through different levels of abstraction. The initial, informal mathematical activity should constitute a concrete basis from which students can abstract and construct increasingly sophisticated mathematical concepts. Students bridge the gaps between concrete and increasingly abstract levels through their creation and use of models, drawings, diagrams, tables, or symbolic notations.

3. Students' learning of mathematics and progress through levels of understanding are promoted through reflection on their own thoughts and those of others. Students must have the opportunity at critical points to reflect on what they have learned and to anticipate where the instructional sequence is heading.

4. Learning takes place not in isolation but rather in sociocultural contexts. Consequently, interaction should be an essential component of instruction. Instructional activities should encourage students to reflect, to explain and justify solutions, to understand other students' solutions, to agree and disagree with one another, and to question alternatives.

5. Mathematical understanding is structured and interconnected. Real phenomena, in which mathematical structures and concepts manifest themselves, usually contain mathematics from multiple disciplinary strands. As a result, learning strands should be intertwined rather than independent.

AN INSTRUCTIONAL SEQUENCE BASED ON RME

Perhaps the most obvious difference between an instructional sequence based on RME and a more traditional sequence is their starting points. In explaining algebraic representation, most traditional algebra texts start with abstract expressions involving variables and rapidly move to formal equations and their manipulation. After attaining some degree of facility in manipulating equations, the student of algebra is invited to apply these skills to context-based problems.

Materials based on RME, however, reverse the progression. Instead of starting in the abstract realm and moving toward the concrete application, the mathematics starts in contexts (principle 1) and gradually progresses to formal symbolism (principle 2). This shift allows students to engage in meaningful, preformal algebraic activity in earlier grades than they traditionally have. Through a structured instructional sequence, students explore and rediscover significant mathematics that anticipates the more formal representations found in traditional algebra.

The remainder of this article will explore the variety of representations that emerge from two algebra units in the middle school curriculum *Mathematics in Context,* which is based on the principles of RME. "Comparing Quantities" (Kindt et al. 1998) is a unit for grade 6, and "Get the Most Out of It" (Roodhardt et al. 1998) is for grade 8. The following representations of linear relationships will be examined: pictures, invented symbols, combination charts, tables, and algebraic symbols. The focus of the discussion will be on the progressive formalization of students' representations and their relationship to standard algebraic notations.

PICTURES

Figure 19.1 shows a problem that "Comparing Quantities" represents in pictures. Although the balanced scales imply an equality of measure that could be represented in equations, this is purposely not done. Since the authors include this problem on page 2 of the unit, they clearly do not expect students to use any formal algebraic procedures to solve it. Instead, they anticipate that students will approach the problem in whatever way they understand. Solutions reveal a variety of representations that preview more formal representations and strategies that will come later in the instructional sequence.

Bananas

2. How many bananas are needed to make the third scale balance? Explain your reasoning.

Fig. 19 .1

Some students solve this problem using pictures, as illustrated in figure 19.2. Other students use words in equations, in the manner shown in figure 19.3.

Fig. 19.2

10 bananas = 2 pineapples, so
5 bananas = 1 pineapple
1 pineapple = 2 bananas + 1 apple
5 bananas = 2 bananas + 1 apple, so
3 bananas = 1 apple

Fig. 19.3

This informal process is very close to a formal solution involving variables and equations. Other students assign an arbitrary weight to the pile of bananas—for example, they might say, "Suppose the bananas weigh 10 pounds," and reason from there, using words.

Figure 19.4 shows a problem that is stated only through pictures. This problem can be solved using a number of different equivalent representations, but for some students the most obvious strategy involves the pictures. They notice that if they cover up (or take away) a cap and an umbrella from each row, then the difference in price ($4.00) is the difference between the cap and the umbrella.

© 2001 Encyclopaedia Britannica, Inc.

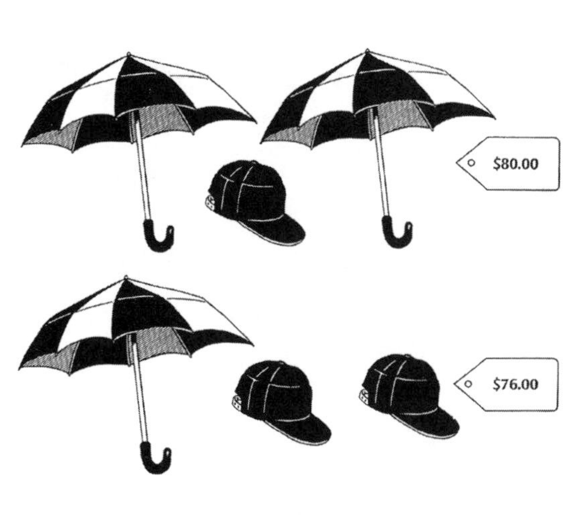

Fig. 19.4

Other students see the two rows of pictures as forming a series that can be continued by producing a third row containing all caps. Figure 19.5 shows the result.

When asked to describe what they have done, the students give explanations such as the following: "When one of the umbrellas in the first row is replaced with a cap, the total goes down by $4.00. So I replaced the umbrella in the second row with a cap and now I have 3 caps totaling $72.00. Now I can find the cost of one cap by dividing." In a similar manner, they might start from the lower row in figure 19.4 and produce a new row above the upper one, containing three umbrellas for a total price of $84.00.

Fig. 19.5

INVENTED SYMBOLS THAT RESEMBLE VARIABLES

When working with the pictures, some students represent a cap with the letter *C* and an umbrella with the letter *U*. This is a shorthand notation that emerges quite naturally. It takes too much effort to draw an umbrella, so they use a *U* instead. They might even write $3U$ to represent *UUU*, but it means three umbrellas—not three times the price of one umbrella, as it would if *U* were being used as a variable. Students even use this invented notation to create "equations" that are in the same form as they would be with variables. Using this notation, they translate the picture "sentences" into the statements shown below.

$$2U + 1C = 80$$
$$1U + 2C = 76$$

This invented notation leads students toward using *U* more abstractly, implicitly representing the cost of an umbrella. The invented notation often leads them to perform operations on the equations that mirror standard operations on equations. For example, students might add the two equations to get

$$3U + 3C = 156.$$

This operation is supported by the context, and the result is meaningful. It is also reasonable within the context to divide this equation by 3 to get

$$1U + 1C = 52.$$

Then, by taking just a few more steps, students can find the cost of one cap. These preformal representations of variables and operations on equations emerge from the context and are prompted by the pictures and a desire to be economical with words, pictures, and symbols. It probably does not occur to the student that these symbols might have power beyond the context. The symbols simply provide a faster way to represent the situation than pictures or words do.

COMBINATION CHARTS

The unit "Comparing Quantities" introduces the use of combination charts as a representational strategy that is quite useful for solving this type of problem. A combination chart would represent the problem of the caps and umbrellas, as shown in figure 19.6.

In general, a combination chart shows totals for all combinations of two values. In this problem, the values are the costs for caps and umbrellas. By exploring many combination charts representing different contexts, students learn that each chart contains many patterns in the numbers. They discover that they can use these patterns to solve problems that they can express as two numbers in a combination chart. For example, students soon see that each combination chart has consistent diagonal patterns. By studying the chart shown in figure 19.6, for example, they learn that when they move down and to the right, the number decreases by 4. They can exploit this pattern to find other numbers on the diagonal and finally to arrive at the cost of three caps when they reach the edge of the chart. Such playing with patterns can be done without any reference to caps and umbrellas. However, in the context, the diagonal pattern of *moving down and to the right decreases the number by 4* is analogous to reducing the number of umbrellas by one while increasing the number of caps by one and lowering the total cost by \$4. Also, finding the number on the edge of the chart where it shows a

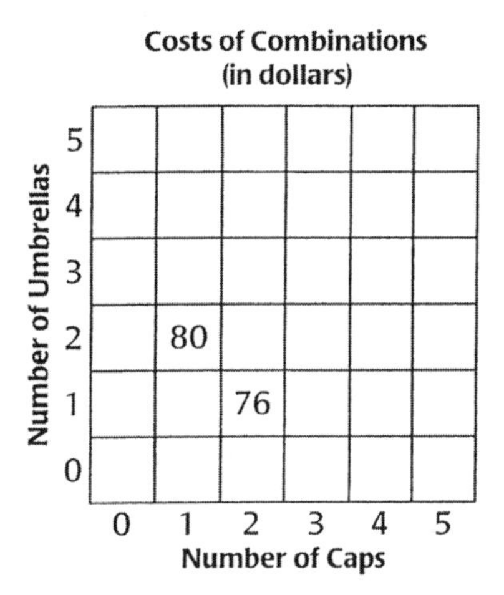

Fig. 19.6

combination of three caps and no umbrellas is analogous to drawing a new picture equation that shows three caps equal to $72.00. By the same token, a picture equation showing that three umbrellas is equal to $84.00 is analogous to moving up and to the left on the chart, with an increase of $4, to find the combination for three umbrellas and no caps. Figure 19.7 shows the two approaches.

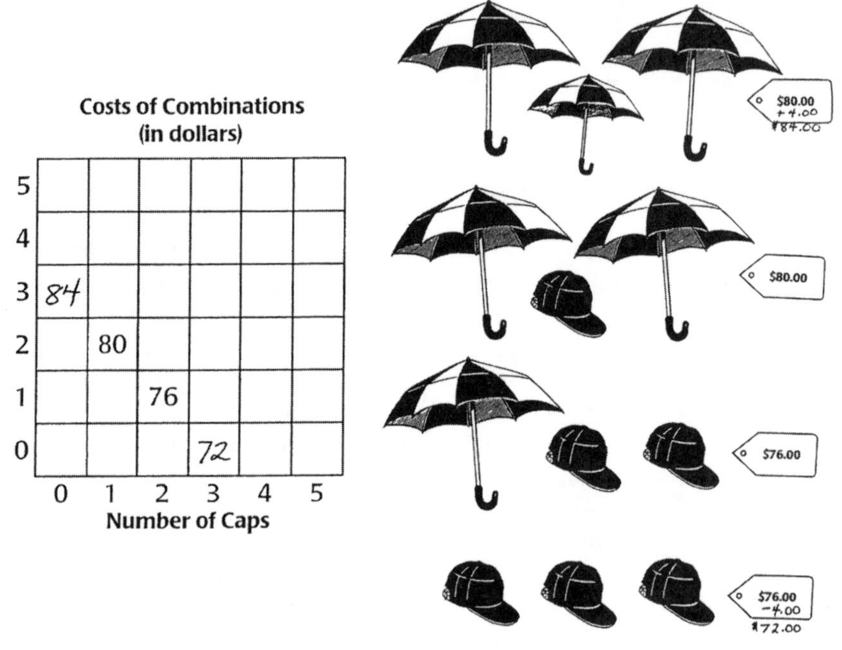

Fig. 19.7

TABLES

"Comparing Quantities" introduces another representational strategy for solving problems of this type. This strategy uses what the unit calls "notebook" notation, since it is introduced in the context of writing down information in a notebook. A waiter is taking orders and tallying costs in a restaurant. Figure 19.8 presents the problem, using both pictures and notation.

Some students solve the problem using the pictorial representations, whereas other students don't see them as containing useful information. The totals for orders 4 and 5 in the notebook can be found by the simple division of previous orders (1 and 3). Total prices for orders 6 and 7 are somewhat more difficult to determine but quickly become apparent when students recognize each of them as the sum of two earlier orders (order 4 + order 5 = order 6, and

Mario's Restaurant

Mario runs a Mexican restaurant, and he is very busy. He moves from one table to another, writing down all the orders. Below you see how he writes the orders on his order pad.

ORDER	TACO	SALAD	DRINK	TOTAL
1	2	4	–	$10
2	1	2	3	$8
3	3	–	3	$9
4	1	2	–	
5	1	–	1	
6	2	2	1	
7	4	2	3	
8				
9				
10				

3. Some of the orders do not have total prices indicated. What are the prices of these orders?

4. Make up two new orders and write them in your notebook. Fill in the prices of these orders.

5. What is the price of each individual item?

Illustrations and text of this and preceding figures in this article are reprinted with permission from the *Mathematics in Context* program © 2001 Encyclopaedia Britannica, Inc.

Fig. 19.8

order 3 + order 4 = order 7). Teachers may see the first three orders, for which prices are given, as a 3×3 matrix and may recognize that the missing totals for the other orders can be found by using the results of elementary row operations (dividing by a scalar, adding rows, etc.). Students are unaware of this matrix format. They see the rows and columns on Mario's pad simply as a shorthand way of writing the information of the problem.

Classroom interaction (principle 4) reveals that students do in fact see the difference between this problem and the one involving the caps and umbrellas. They recognize that there are *three* prices to find here instead of just two. Students often ask if a combination chart could be used to represent the combinations of information about tacos, drinks, and salads. Some even try to imagine a three-dimensional equivalent to the combination chart but quickly realize that such a thing would be impractical, even if it were theoretically possible. This reflection on the instructional sequence illustrates principle 3 of RME.

EQUATIONS

The instructional sequence described above concludes with a section in which variables and equations are introduced more formally by relating them to combination charts and the notebook notation. The equations are always related to a familiar context that supports their meaning, and as a result, equations are seen as just one more way to represent information. The fact that equations can be explored for their own sake, apart from context, is something that comes in later study. At this point in grade 6, a preformal use of equations with understanding is the goal of instruction.

The expanded notion of equation as a representation is seen in the following problem from the grade 8 unit "Get the Most Out of It":

> Eighth-grade students from Wingra Middle School are going on a camping trip. There will be 96 people going, including the students and teachers. All the luggage, gear, and supplies are already packed into 64 equal-sized boxes. Now the organizers want to rent the right number of vehicles to take everyone to the campsite. They can choose between two different types of vehicles from a car rental agency:
>
> Minivan: seats 6 people
Cargo space: 5 boxes
>
> Van: seats 8 people
Cargo space: 4 boxes

A formal algebraic approach would focus on the simultaneous solution of the two equations that describe the situation: $6M + 8V = 96$ and $5M + 4V = 64$. However, in the preformal approach adopted by RME, students use the representations they have learned so far and clues from the context.

Questions follow that encourage students to explore different combinations of vans and minivans and the resulting numbers of people and packages that can be transported, with the idea of eventually finding the smallest number of minivans and vans that are needed. These explorations are all performed by examining the meanings supplied by the context. For example, the unit suggests that if students consider just the people, they will see that 4 minivans can be exchanged evenly for 3 vans. "Why does this work?" the unit asks the students, and then it directs them as follows: "List all the possible combinations of minivans and vans that carry exactly 96 people." This effort might result in a table such as that shown in figure 19.9. Similar questions related to an even exchange between minivans and vans for the boxes and all possible combinations will lead students to the solution that 8 minivans and 6 vans make up the smallest number of vans and minivans needed to carry 96 people and 64 boxes.

Minivans	Vans
16	0
12	3
8	6
4	9
0	12

Fig. 19.9

The next strategy that students encounter is a graphical one. Using the table in figure 19.9 and the one resulting from the students' examination of the combinations of minivans and vans needed to carry exactly 64 boxes, students produce the graphical representation of the problem shown in figure 19.10. A specific question directs students to discover that in the context of the problem, most of the points on either line are not solutions, because you can rent only whole vehicles.

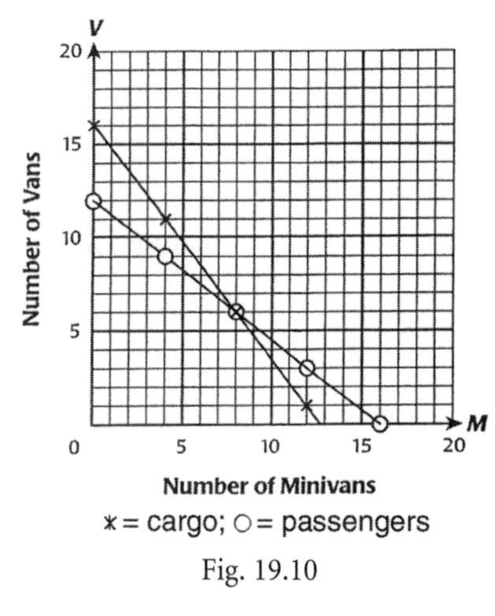

Fig. 19.10

Students can also solve the van and minivan problem using combination charts. The chart in figure 19.11a is designed to show the number of people carried by various combinations of minivans and vans. The number 96 is written in the appropriate cells because the problem requires that exactly that many people be transported. The chart in figure 19.11b is designed to show the number of boxes carried by various combinations of minivans and vans. Here, the number 64 is written in, because the problem requires that that many boxes be transported. The similarity between the combination charts and the graphical solution is of course no accident. The representations reflect the same relationships between minivans and vans in the numbers of boxes and people that they can carry. In figure 19.10, the even exchange between minivans and vans is reflected by the slopes of the two lines. The line that represents people reflects the exchange of four minivans for three vans by the slope of $-3/4$ for the graph of $6M + 8V = 96$. In the combination chart, the combinations totaling 96 are 3 down and 4 to the right of each other (or 3 up and 4 to the left).

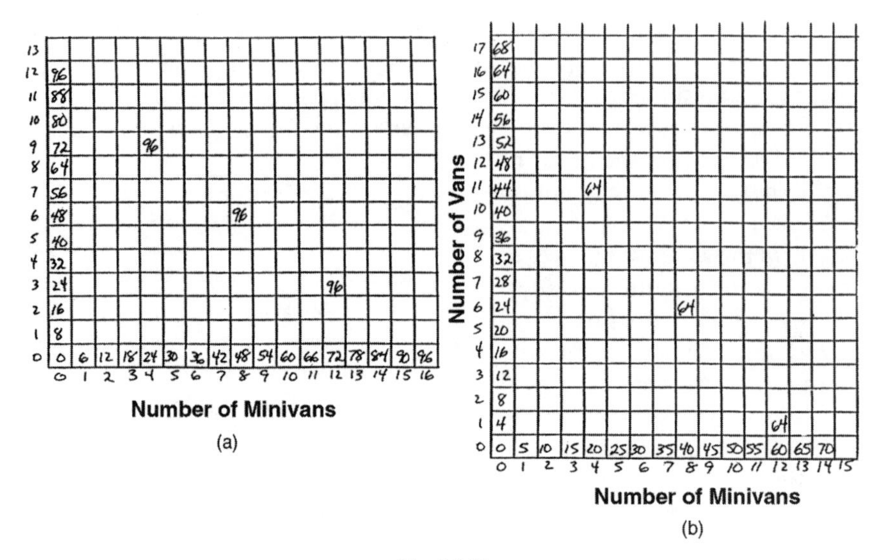

Fig. 19.11

The final strategy presented in the context of this problem involves the following two equations:

$$6M + 8V = 96$$
$$5M + 4V = 64$$

The numbers and letters are clearly attached to the context, and students can express their meaning. For example, in the equation $6M + 8V = 96$,

6 stands for six people in each minivan.
M stands for the number of minivans.
8 stands for the number of people in each van.
V stands for the number of vans.
96 stands for the total number of people who must be carried.

The mathematical structure of this set of equations is identical to that of the set of equations for the caps-and-umbrellas problem. The linear equations for both problems represent the context. However, an interesting thing happens when students apply one of the strategies that the sixth-grade unit used for solving that earlier problem, as shown:

$$\begin{array}{lcrcr} (5M + 4V = 64) \times 2 & \longrightarrow & 10M + 8V & = & 128 \\ 6M + 8V = 96 & \longrightarrow & \underline{6M + 8V} & \underline{=} & \underline{96} \\ & & 4M & = & 32 \\ & & M & = & 8 \end{array}$$

Substituting the value for M in the second equation results in $V = 6$, a value that students can check against the previous solutions. Some students may object that the equation $10M + 8V = 128$ that resulted from doubling the first equation appears to have changed meaning within the context. They may point out that the minivan can still hold only 5 boxes—not 10—and that the van can still hold only 4 boxes—not 8—as suggested by the new equation. A subtle rewriting can restore their concrete understanding of meaning in the context. When the equation is written $5(2M) + 4(2V) = 128$, these students should see that it is the number of *vehicles* that has been doubled—not the number of *boxes* that the vehicles can hold. Now, the operation on the equation appears reasonable, both from a mathematical point of view and in the context. Most students at this level would never notice any problem with the doubled equation. They are not bound by the context, and they accomplish a transition to the abstract without confusion.

The Five Principles Revisited

We have explored a variety of representations that emerge from a sequence of linear problems that are based on concrete situations, in keeping with the tenets of RME. The first principle of RME was manifest in the way in which the contexts stimulated these representations. These concrete contexts allowed students to engage in meaningful mathematical activity from which their understanding could grow. The second principle of RME was embodied in the progressive levels of abstraction reflected by the pictures, charts, tables, and equations presented in the units and in students' solutions to problems. Teachers should note that even after students have encountered formal equations, it is not unusual for them to operate with preformal models. It is always preferable for a student to use a less formal strategy with understanding than a more formal one without understanding. The third and fourth principles are most evident in classroom interactions, and as a result, they are difficult to illustrate here. It is clear, however, that reflection and interaction can enhance students' understanding and appreciation of many of the representations shown in this article. Finally, the fifth principle is illustrated by the connections between the two units.

The reader might wonder about the learning outcomes of the instructional sequence described here. At a minimum, the instructional sequence appears to allow students to engage in significant mathematics with understanding at earlier grade levels than more traditional programs do. It would also seem reasonable to conjecture that students gain from this sequence a deeper understanding than students ordinarily achieve of the connections among representations, in whatever form, and the context they reflect. A related outcome is also possible—that is, that students can model contexts with symbols, which they in turn can use to solve problems related to those contexts. This expectation that mathematics is about "making sense" is often suspended or forgotten when students move to abstract representations too quickly and without the grounding provided by realistic contexts. Another possible outcome of the instruction is an understanding that situations presented in problems can often be represented in different ways and that the different representations suggest different strategies for solutions. These learning outcomes are currently only conjectures, if not high hopes. Nevertheless, they are plausible, significant, and worthy of investigation.

REFERENCES

de Lange, Jan. *Mathematics Insight and Meaning.* Utrecht, Netherlands: Vatgroep Onderzoek Wiskundeonderwijs en Onderwijscomputercentrum, 1987.

———. "Higher Order (Un-)Teaching." In *Developments in School Mathematics Education around the World: Proceedings of the Third UCSMP International Conference on Mathematics Education,* edited by Izaak Wirsup and Robert Streit, pp. 49–72. Reston, Va.: National Council of Teachers of Mathematics, 1992.

Kindt, Martin, Mieke Abels, Margaret R. Meyer, and Margaret A. Pligge. "Comparing Quantities." In *Mathematics in Context,* edited by the National Center for Research in Mathematical Sciences Education and the Freudenthal Institute. Chicago: Encyclopaedia Britannica Educational Corporation, 1998.

Roodhardt, Anton, Martin Kindt, Margaret A. Pligge, and Aaron Simon. "Get the Most Out of It." In *Mathematics in Context,* edited by the National Center for Research in Mathematical Sciences Education and the Freudenthal Institute. Chicago: Encyclopaedia Britannica Educational Corporation, 1998.

Treffers, Adri. "Didactical Background of a Mathematics Program for Primary Children." In *Realistic Mathematics Education in Primary School,* edited by Leen Streefland, pp. 21–56. Utrecht, Netherlands: Freudenthal Institute, 1991.

20

Charting a Visual Course to the Concept of Function

Michal Yerushalmy

Beba Shternberg

MODELING, which is the representing of authentic situations, has been integrated into many recent approaches to teaching. Modeling can be considered a form of mathematical challenge that supports the emergence of representations of functions. Although mathematical representations of situations have always been a part of the mathematics curriculum, graphing technology (such as that suggested by the National Council of Teachers of Mathematics (NCTM) Standards for algebra and technology [Heid 1995]) allows for forms of modeling that differ substantially from pretechnology environments. Research suggests that mature solvers of word problems use representations of functions (by which we mean real-valued functions of a real variable) as modeling tools. This use, however, has never been the intent of traditional work with word problems in algebra (Hall 1989). In traditional curricula, students were presented with "word problems"—verbal descriptions of situations that were carefully crafted to contain all the information necessary for the students to make a list of data "given" and "to be found" before they could solve the problem. So, solving problems that are connected to situations was an "application" activity, where students applied a specific, known formula or method of symbolic calculation.

There is evidence that algebra students for whom the concept of function is central are able to identify the important properties of a situation, as displayed in any representation, and are better equipped than other algebra students to solve traditional algebraic word problems (Hershkowitz and Schwarz 1999; Heid 1996). One explanation for this is that the use of multiple-representation technology increases students' opportunities to operate with math-

The writing of this article was supported by a grant from MISES: the Study of Educational Systems, Milken Institute, Jerusalem.

ematical representations of situations and to analyze them in a connected fashion.

However, writing expressions and equations is still considered to be a complicated task for algebra beginners (MacGregor and Stacey 1999). Although graphs are central to representing situations in algebra that can be modeled as functions and although links among alternative representations often provide new insights that support reasoning processes, the visual representation nevertheless has emerged as a weaker source for experimentation than symbolic representations. This is perhaps because graphing technology preserves the symbolic representation as the major channel into the mathematical model. Thus, students typically experiment with representations only when they are already somewhat knowledgeable about the symbolic language of algebra. In figure 20.1, we suggest that in many current curriculum sequences, the links between phenomena and mathematical representations of functions are introduced only after students have been taught to think of functions as interpretations of algebraic expressions.

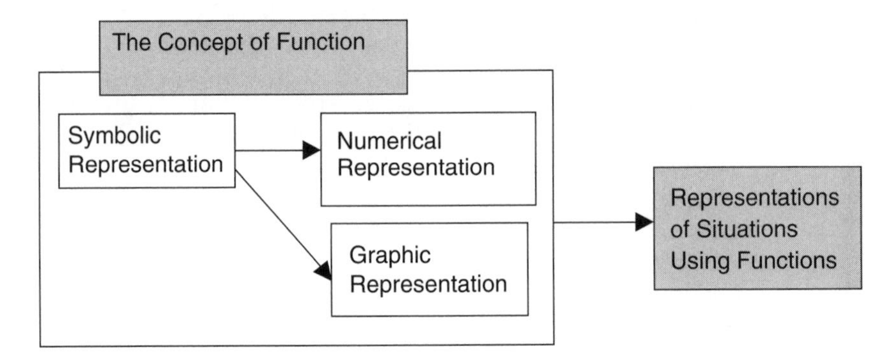

Fig. 20.1. Symbolic representations for functions as input for graphing technology

Current work with technology, however, supports another type of experimentation with situations and representations. Microcomputer-based laboratories (MBLs) allow students to represent an ongoing situation using devices that gather data from it and present the data graphically. These technologies eliminate algebraic symbols as the sole channel into mathematical representation and motivate students to experiment with the situation—to analyze and reflect on it—even when it is too complicated for them to approach symbolically. The visual analysis elicited by work with MBL tools is different from visual analyses that arise from work with conventional representations of algebraic symbols or traditional numerical tables. This difference may be attributable to the continuous nature of the MBL graphing, which calls attention to the geometric properties of the graph and to the ways it relates to the qualities of the situation (Nemirovsky 1994). But how does this visual approach to functions support students' symbolic understanding? One answer is that it shows them important connections by moving back and forth between measurements and machine-made symbolic expressions. Measurements can be presented by MBL tools in numerical tables (spreadsheets or spreadsheet-like tables), and data-fitting tools can then construct symbolic expressions that match the graphs and the numerical data.

We conjecture that although this sequence provides a technological solution that supports a complete, multiply-linked representation system, the links established among the representations are too weak to allow beginners to build a conceptual understanding of the mathematical symbols. Although using the MBL technology supports visual investigations of functions and graphs and in later stages of the learning may support even solid connections between symbolic expressions and phenomena, additional efforts should be made to explore how to *introduce* symbols in a way that would allow students to benefit from visual experimentation with situations (see fig. 20.2).

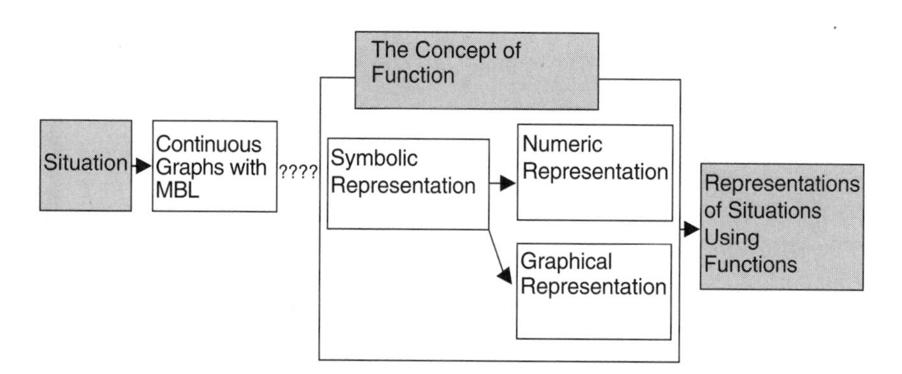

Fig. 20.2. Modeling situations before using symbolic representations for functions

The symbolic representation of functions in algebra is the focus of many educational discussions (Arcavi 1994), which typically focus on the difficulty of predicting changes that result in the graph of a function from algebraic transformations of the expression that defines it. We have therefore sought ways of supporting students in developing a mathematical understanding of all representations, including symbolic ones, by making the connections between expressions and graphs as transparent as possible.

In our work, we follow Nesher (1989), who suggests that any systematic learning of a new concept should be done within an environment that is intentionally designed to support a series of transitions, from working with familiar objects, to carrying out mathematical processes using representations of these objects, to encapsulation (constructing a cognitive mathematical object) of the process, to a new object that then can be manipulated. We have tried to design learning environments that would bridge the formal language of mathematics and the informal experiences of students, and we have tried to give students opportunities to use their creative resources to construct the concept of function. In this paper, we will outline the essential parts of a curriculum sequence, which is designed to support the construction of the concept of function as the foundation for algebra.

Using episodes from our work with seventh graders in the Visual Mathematics curriculum (1993–96), we will describe three major learning phases that students go through before learning algebra. The first phase involves them in building a graphic representation of a situation. It begins with students drawing trajectories of an object in motion and then focuses on the transition from drawing mathematical graphs to analyzing them as representations of functions. In episode 1 below, which exemplifies the first phase, the goal is for students to describe an authentic situation (the motion of a thrown ball) by sketching it on a plane. Students study the graphs of the process as it changes over time, and they develop narratives to connect the actions of the situation with the features of the graphs.

The second phase is intended to support the students' transition from a graphical representation of a specific situation to a system of iconic representations of functions. It supports the encapsulation of the process by providing a set of visual constraints in the form of icons that limit verbal descriptions to those strong enough to describe any well-behaved function's graph and the associated process that such functions describe. In episode 2, we glance at a problem that focuses on an analysis of rates—a topic traditionally addressed in calculus as an application of taking derivatives. Using this problem—which for most problem solvers would be too complex to be described symbolically—we demonstrate the potential of visual objects to become productive tools that support modeling and a meaningful resource for approaching formal operations with functions.

The third phase leads to another transition—from the iconic representations of the second phase to the symbolic representation of functions. Episode 3, which illustrates this phase, shows a construction by students of symbolic representations. The students created this construction by manipulating the graphic representation and then exploiting the connections between its rate of change and its accumulation.

By describing the three phases of the designed sequence, we hope to demonstrate a curricular route that can support prealgebra students as they develop the mathematical idea of function in all its representations.

PHASE 1, EPISODE 1: FROM AUTHENTIC SITUATION TO VISUAL REPRESENTATION

Phase 1 focuses on the transition from drawing a situation to analyzing mathematical graphs as representations of functions. Episode 1 illustrates the first phase, in which students build a graphical representation of a situation.

> Alon, Erez, Maya, and Hila were trying to represent the following situation:
>
> "Hava loves to play with the ball. She throws it to the ground. The ball hits the ground and then it hits the wall."

Drawing freehand with the mouse, they used a component of the Function Sketcher software (Yerushalmy and Shternberg 1993) to plot the path of the ball on a Cartesian system. The software recorded the position coordinates and provided (simultaneously or on request) graphs of the parameters of the motion over time (e.g., x-coordinate, y-coordinate, the accumulated length of the path, the distance from a given point). Presumably because the drawing was dynamic and the appearance of various graphs simultaneous, the students did not fall into the common trap of confusing drawings and graphs. On the contrary, they were faced with the challenge of explaining the differences between the drawing and the graphs, an activity that led them to elaborate on the meaning of the graphic representation. As the students drew the ball's path on the basis of their personal experience, they were discussing their view of the motion of the ball at the same time that they were receiving graphs of the path's x- and y-coordinates over time (see fig. 20.3).

Alon and Hila looked first at the $y(t)$ graph. Probably because the graph resembled the path and because the vertical coordinate matched the height of the ball, they had no trouble matching parts of the graph with parts of the phenomenon (see their comments on the $y(t)$ graphs). The $x(t)$ graph, however, is less geometrically intuitive, and when they turned to analyze it, Alon ventured, "Oh, that's hard to understand." Indeed, it did take some time for the group to come up with a correct interpretation of the $x(t)$ graph. However, the ultimate goal was not to have the students interpret the graph but rather to have it serve as a trigger, moving them easily between graphs as representations of functions and graphs as representations of situations.

Soon afterward, Hila challenged the group with additional questions about the graph of the x-coordinate in time, which continued to puzzle her. She gradually began to analyze the role of the person behind the behavior unfolding in a temporal process. As the group traced the ball's path using the mouse, Hila perceived that she controlled the coordinates x and y; both depended only on her drawing decisions. But when she turned to the graphs,

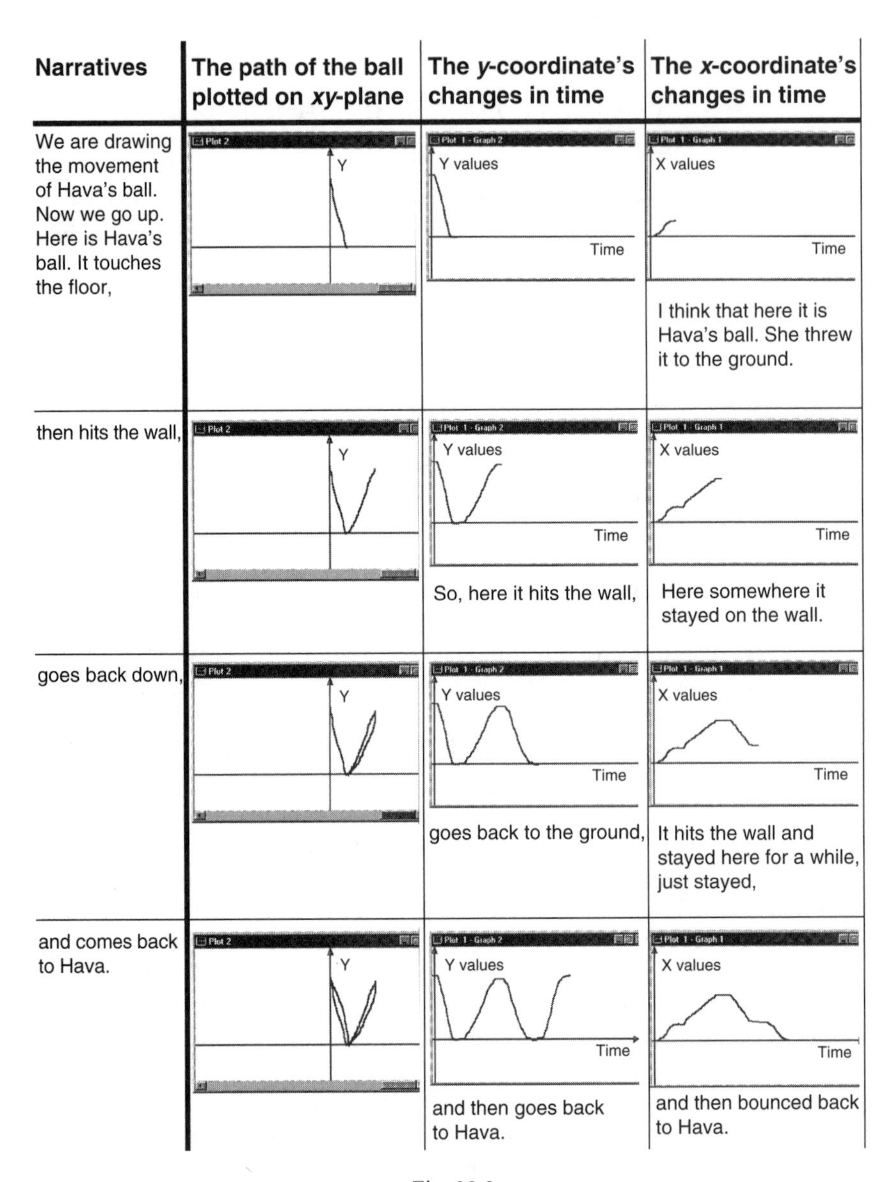

Fig. 20.3

the variable *time*—which was recorded but did not appear in her drawing—now appeared as an independent variable—a variable that seemed to change independently of her actions. She put it this way: "For instance, time. It passes. We can't stop it. Or, for example, our growth. We can't stop it. These are things that don't depend on us.... In a coordinate system, time is ... time is a name of an axis. If your *x*-axis is the time axis, the time just passes by itself."

Different interpretations of Hila's remarks could be relevant here: Does Hila mean that if time is an independent variable, it should have a *causal* relationship to the dependent variable x? Is Hila accustomed to situations where the value of the independent variable is "rhythmical" and discrete (e.g., "He went 5 km in each hour.") rather than continuous?

We were and are interested in Alon's answer, particularly the way in which he formulated it. Alon used the $x(t)$ graph to highlight and resolve the differences between two views of independence in the meaning of *independent variable.* In one view, x is an independent variable because it is *unrestricted*—that is, it is not determined from the outset or by a predetermined rule. In the second view, *time* is an independent variable because it changes without our control. Alon justified the first view by moving the mouse freely among values of x to show that the "hand controlled" quantity is still dependent in time. He then invented an action to show the second, "uncontrolled," view of independence. He fixed the mouse at a point and created a constant graph of x over time (see fig. 20.4).

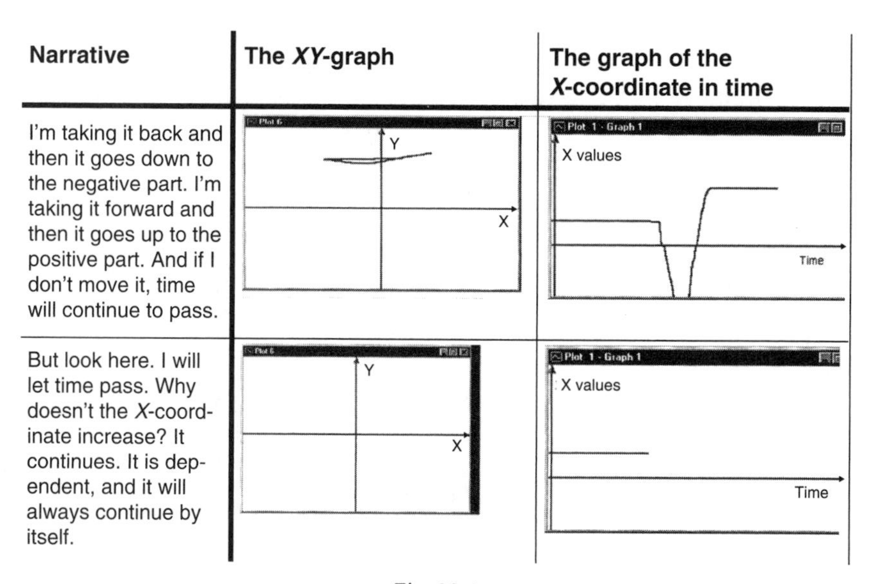

Fig. 20.4

Alon's language became more general, describing three *typical* processes rather than a *specific* motion: x goes up (when he moves right), goes down (when he moves left), or remains constant. His shifts between the action and the graphs seemed to take a different direction from the ones the students were making at the beginning of their work. Rather than merely watching graphs as they resulted from his drawing, Alon *planned* the action that would produce a specific graph. (For example, to create the constant graph, he drew a steady point.) Thus, for Alon and his classmates, graphs typically became *motion*. The students also explored ideas, such as dependency and constancy, that are regarded as central to the understanding of a function (see Hazzan and Goldenberg [1997] for similar ideas that were raised while advanced college students were experimenting with interactive geometry). All this suggested that the students had developed a solid base for making the transition to the more advanced ideas of the next phase of our prealgebra curriculum.

PHASE 2: FROM GRAPHING TO ICONIC VISUAL REPRESENTATION

We designed our second phase to support the encapsulation into objects of the process of graphing. We accomplished this goal by providing a limited set of graphical icons, which imposed visual constraints on the student, and a limited verbal list of function properties (see fig. 20.5). There were synonymous relationships among the iconic and verbal elements of this linguistic representation. Even though statements articulated in one lexicon could be restated in the other, the two distinct sets allowed for different degrees of precision. For example, describing a function in a region as "increasing" and "curved," or even more accurately as "increasing" at an "increasing rate of change," is far less limiting than the display of a *particular* increasing and curved function is.

Embedding the two lexicons in a software environment makes it possible

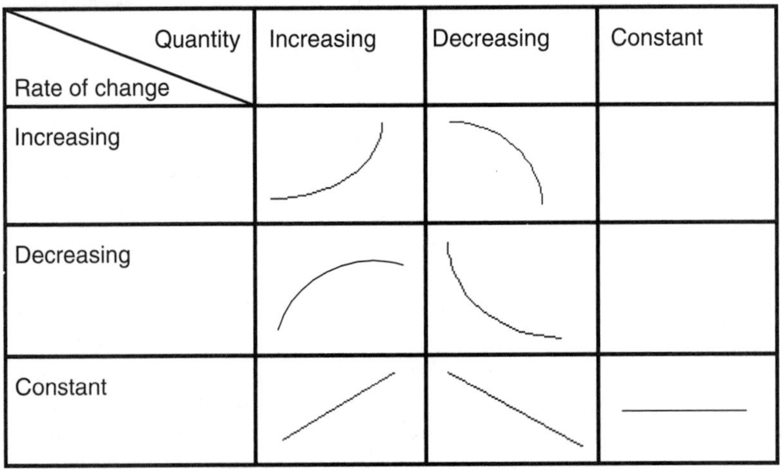

Quantity / Rate of change	Increasing	Decreasing	Constant
Increasing			
Decreasing			
Constant			

Fig. 20.5. Two lexicons to represent the visual properties of functions

for students to express themselves in either sort of lexical element while the software displays equivalent synonymous statements in the other lexicon. Students become familiar with this limited sign set as they use it to represent situations. They sketch graphs for given situations by adjusting the seven icons and creating narratives using the three verbs. Figure 20.6 shows a typical representation. The limited lexicon allows the terms *constant, increasing,* and *decreasing* to be illustrated by more than one icon.

To allow for a finer classification within a verbal description of a particular behavior, the software permits students to introduce "stairs," discrete companions to a continuous graph (fig. 20.7). By adjusting the stair feature, students

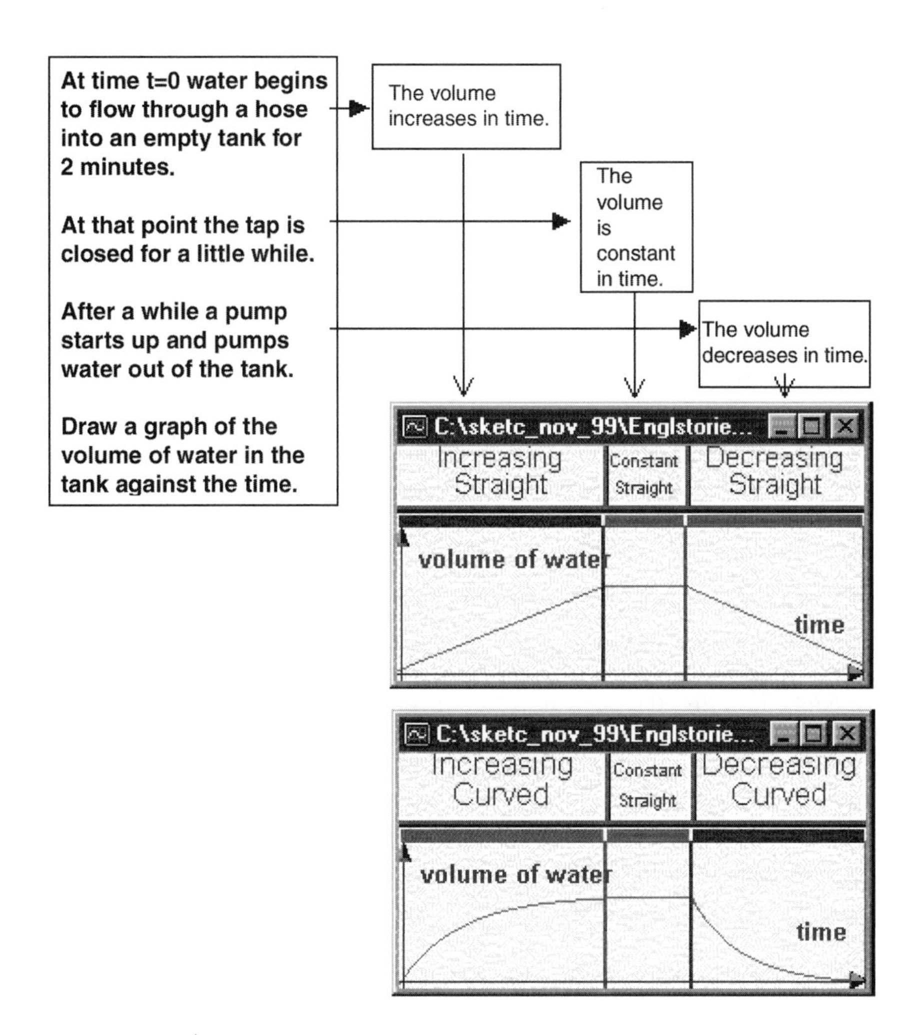

Fig. 20.6. A story text, a list of constraints, and two possible graphs

dents can create many different variations of graphs that satisfy the same set of constraints.

Students eventually adopt the different components of the lexical system (icons, verbs, and stairs) as manipulable objects that support them in solving problems that are too complicated for them to describe symbolically. Episode 2 illustrates typical work with these iconic objects and terms.

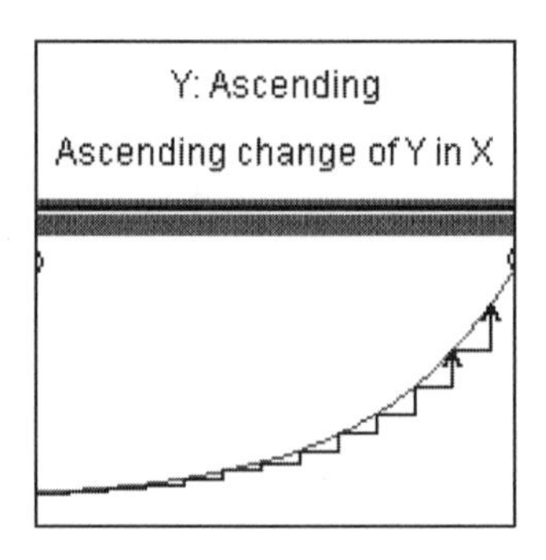

Fig. 20.7. Stair companion of continuous graphs

Episode 2

In earlier work, we had interviewed thirty-four advanced calculus students in the eleventh and twelfth grades. The students were attempting to solve the following problem from a qualitative calculus unit (Taylor 1992):

> A cook has a large portion of meat at room temperature that should be cooked as quickly as possible. He has at his disposal a conventional oven and a microwave. In the microwave oven meat temperature increases at a constant rate, and in the conventional oven it increases at a changing rate. Using the results of an earlier cooking trial, the cook knows that although meat temperature in the conventional oven is always higher than in the microwave, cooking time for this piece of meat would be 2 hours in either one of the ovens. Could cooking time be less than 2 hours using some combination of these two ovens?

The students had taken algebra, precalculus, and calculus courses. However, none of their courses had focused on modeling or authentic situations. Most of the interviewees tried to represent the situation symbolically but failed, since the problem did not include any specific data. In general, our interviewees had very little success with the problem.

For a comparison, we suggested this problem to our seventh graders. Two students, Erez and Ilanit, volunteered to work with us on the problem outside the class. They had already represented situations using icons and stairs, but they had never before worked on a problem as subtle as this. Unlike most of the calculus students, they approached the problem by exploring iconic sketches that would match its conditions.

From the beginning, they decided to use graphs not only to examine the dependence of one variable on another but also to describe and compare rates:

> *Erez:* Good, we do have some facts that we can use to describe the graph we need. The first fact we know is that the microwave raises the temperature at a constant rate.

Ilanit: And the conventional oven heats at a nonconstant rate.

Erez: Fine, so the first thing we know is that if it is constant, then it is straight.

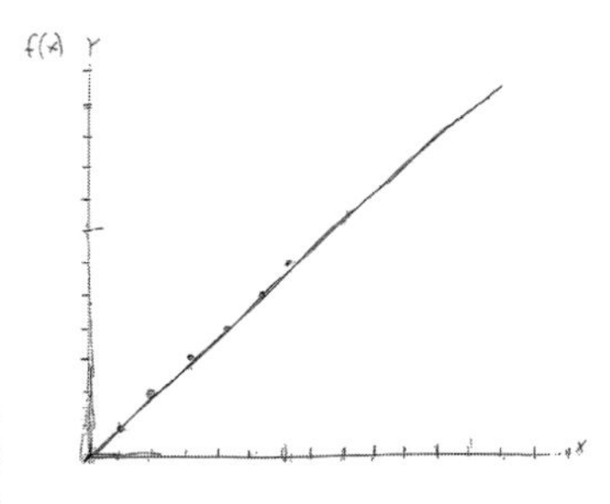

The students analyzed the relevant data, distinguished between quantity and its rate of change, and inferred that one of the graphs had to be straight. They continued by discussing the rest of the data:

Ilanit: We don't have the exact data, but we can conjecture ... [that] the process took two hours.

Erez: The graph continues....

Ilanit: No, the process ends, but mathematically, the graph continues.

Erez: So at the end of the two hours, we'll cut it. Let's say that here we will mark the final point. [*Draws a dotted vertical line to mark the duration of the processes, and erases the rest of the straight increasing line.*]

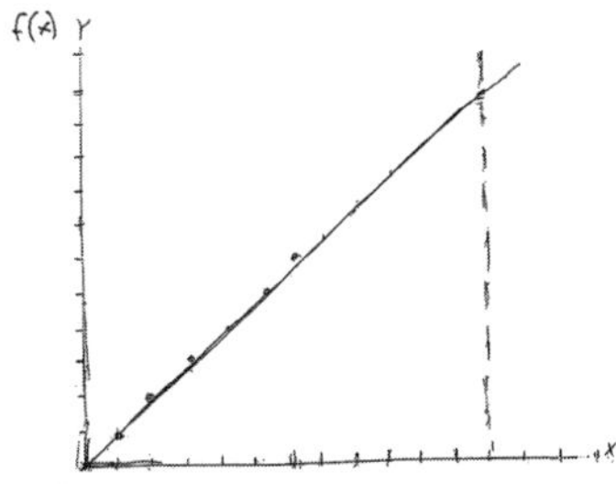

What remained to be determined was the shape of the nonlinear graph, and they had at their disposal exactly two types of graphs to represent processes that increase at a changing rate. The students were ready to draw the second curve after another round of checking:

Ilanit: Yes, something else that we should pay attention to is that along the way the oven will have higher temperatures than the micro.... We should ensure that there will be different temperatures ... no matter how it is going to increase.... Well, it has to be concave down.

Erez: This line ... to describe the givens in the problem, it has to be concave down! Why? Because if it were concave up, it would go below the microwave values.

Once they agreed on the shapes of the graphs, all the given information was fully represented, and Erez and Ilanit moved to solve the problem. Eventually, Erez made a breakthrough, discovering that the stair tool could help them solve the problem:

> *Erez:* At this point, they reach the same point.... The conventional oven leads all the way over the microwave, [and] in a specific position, they ... I got it ... *stairs*!

Using the stair analysis, they shifted back and forth between the graph to the phenomenon as they completed the solution of the problem:

> *Erez:* So it will go this way: it starts this way, and then it [the stair height] gets smaller and smaller, and here [*draws a vertical line*] ... up to here, it will be the conventional oven, and from here, we'll get taller stairs ... here, it starts to be weaker....
>
> *Ilanit:* I think it weakens after an hour sharp!

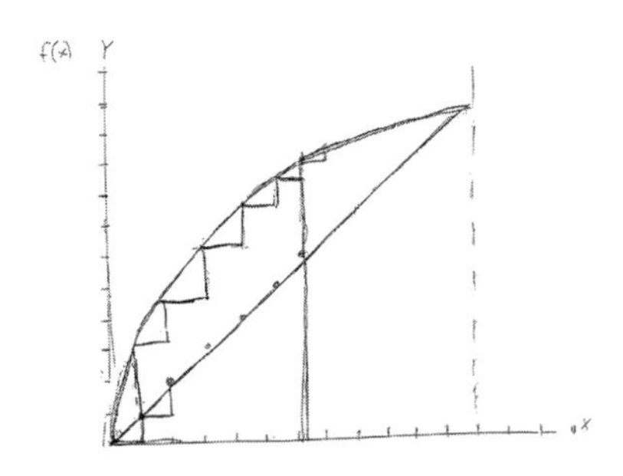

Their last effort was to combine the two graphs into a single graph that would model the whole process and demonstrate the reduced cooking time. By estimating the height of the stairs on the nonlinear graph and comparing them to the stairs of constant height, they determined the point where the cook should move the meat from one oven to another. Working with the information they had learned, they ultimately drew a single continuous graph to represent the solution:

> *Ilanit:* Let's say it's an hour here and two hours here.... It will look curved in the first part and straight in the second. I'll sketch it.... An hour here, curved, and then straight.

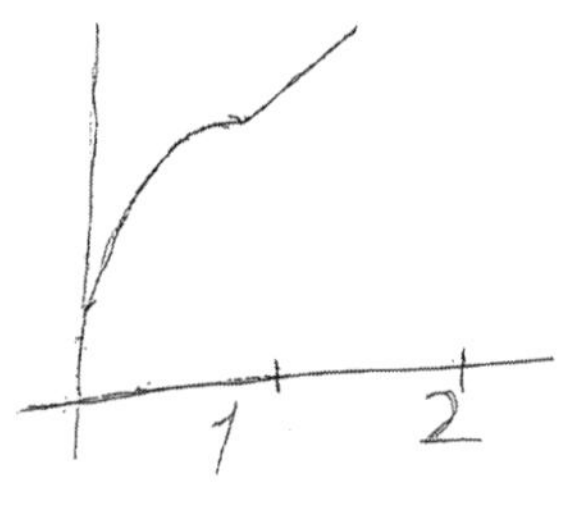

Ilanit and Erez solved the problem using a limited set of graphical terms that functioned as objects for describing processes. The visual objects (iconic graphs and stairs) represented the reality of the processes but were manipulated mathematically.

The seventh graders, unlike the calculus students in our earlier sample, immediately resorted to a graphical representation of the situation as a basic tool for describing and analyzing it. Focusing on rate as the key for explaining the behavior of a phenomenon, the students were confident about taking a nonspecific, graphical approach to a solution.

In our view, they had progressed a great distance from drawing situations that were graphed by the computer, through encapsulating processes, to visually representing them using an iconic language, to using the full expressive potential of the language by employing the stair representation. Next we worked to expand students' skills in using the visual system of objects and manipulating those objects, as the students became increasingly fluent in the use of symbolic representations of situations.

PHASE 3: VISUAL REPRESENTATION OF RATE OF CHANGE

In this last phase of our prealgebra curriculum, students see how the components of their graphical representations can lead to the construction of formal symbolic representations for functions. We begin by experimenting with symbolic notations for constant-rate processes. In part, this process reflects the traditional algebra curriculum at this stage, but we have also discovered that our students are already using the stair representation to compute the value of a function for any integer input. Students often use it for a function f to compute $f(n)$ by summing the heights of the "risers" on the stairs that bring one to a height $f(n)$. In essence, they were accumulating the sum of the first differences. If the risers all have the same height a, the repeated addition becomes multiplication, and students arrive at the formula $f(n) = an$, which they interpolate to produce a linear function defined on the real numbers given by the same formula (see fig. 20.8).

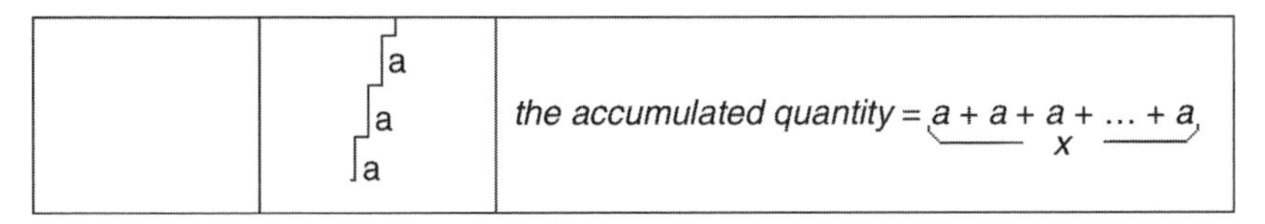

Fig. 20.8. Students' stair sketch and its expression to a line

Episode 3

We presented a problem that is an old favorite:

> A king was defeated in a chess game on the 80th day of the competition by a poor stranger. As promised, the stranger got the prize – coins of gold, consisting of one

coin for the first day, two for the second, three for the third, and so on, with each day's coins amounting to one more coin than the previous day's. How many coins did the stranger receive?

Because of their experience, our students did not see the problem as presenting an arithmetic series (as it usually is treated in the textbooks), but rather, they inferred that the quantity of coins was increasing at an increasing rate, and they sketched an appropriate graph using stairs. Connecting the graph and the stairs to the numerical patterns of rate of change, they agreed on the meaning of the width of each stair (a single day) and the height of each stair (the appropriate number of coins). They then looked for methods to find the sum of coins accumulated in 80 days.

Anat drew a representation of her group's work on this problem and explained her and her colleagues' method to her classmates. While she was talking, she began trying to use the strategy demonstrated in figure 20.8, and she drew an alternative set of stairs of equal heights and wrote a numerical expression for the amount of coins. Following Anat's presentation, the class tried to generalize the strategy and find an expression to describe the nonlinear phenomenon. With the help of their teacher, the group replaced the 80 days with a variable and created a step-by-step translation of the stairs representation into symbols. Anat's explanations and the generalizations appear in figure 20.9.

The approach that Anat's group used was similar to that illustrated in figure 20.8 for the linear process. The students matched the value of each stair's height and the drawing of the stairs to draw a curve increasing at an increasing rate. However, turning this process into a symbolic expression was not as easy as it was in the linear example. These seventh graders had not yet worked with sums of sequences, but some of them were able to manipulate the stairs in a way that turned this problem into an equivalent linear problem.

Thus, the phenomenon in the story invoked an iconic graphic representation using stairs. By accumulating stair height and then replacing the stairs with a "constant rate" staircase in which the stair height was the average of all the increasing heights, students built a model that allowed them to find the value $f(t)$ for any number t of stairs. On the basis of Anat's description, a conventional algebraic notation

$$(1 + t) \cdot \frac{t}{2}$$

was constructed. (The students were fortunate to be working with an even number of stairs. They would have had to modify their method slightly if the chess game had gone on for, say, 83 days.) As in their work with linear functions, the students discovered an opportunity to build a symbolic representation of function, based on its graphical representation. We acknowledge that such constructions are limited to processes defined on discrete domains. However, we conjecture that symbolic constructions that are associated with visual properties and are based on typical types of rate of change (such as linear, quadratic, and exponential) constitute an important stage in the formation of the concept of function. This conjecture is grounded in our experience, as well as on evidence from work with calculus students (Gravemeijer and Doorman 1999) and an analysis of the development of algebraic representations (Dennis and Confrey 1996).

Anat's narrative and graphs

Imagine a "stairs" graph that represents this story. To find the value of the height at some point, we can compose all the heights of the stairs that "lift us" to the point. Let's say that the height of the first stair is 1 and every next stair is bigger by 1 from the previous. It means that we have to add:

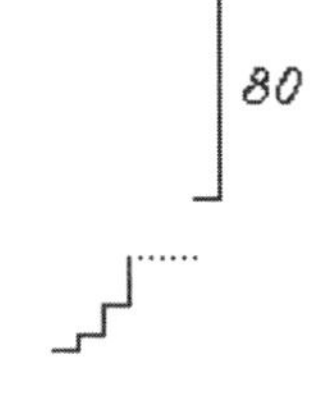

1 + 2 + 3 + 4 ... till the last one (for example, till 80).

Now we add the height of the first stair with the height of the last one, the height of the second stair with that of the one before the end. We get equal sums. It means that we could replace our "staircase" by another one.

$1+2+...+79+80$

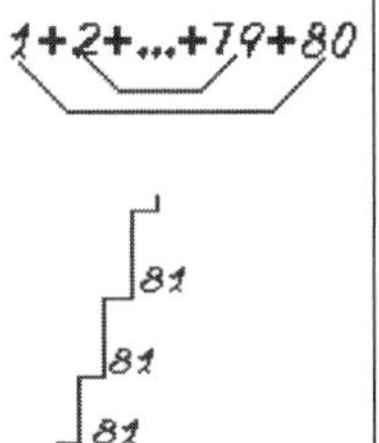

The new staircase has equal stairs [stairs of equal height], and their number is twice less than [one-half] the number of stairs in "our" staircase. The new staircase lifts us to the same height in the end. We know how to compute [the sum of] its stairs' heights.

$(1+80)\cdot\frac{80}{2}$

Later generalizations to symbols

The height of the first stair is 1.

Every next stair is bigger by 1 from the previous one.

The number of stairs is t, which means that the last stair height is also t, and the sum of all heights is

$1 + 2 + 3 + \ldots + t = ?$

The sum of all heights is

$1 + 2 + 3 + \ldots + t =$
$(1 + t) + (2 + (t-1)) + (3 + (t-2)) + \ldots$

Every sum in parentheses is $1 + t$, and there are $t/2$ number of parentheses. The sum is

$$(1 + t) \cdot \frac{t}{2}.$$

Visual Representations as a Means to Form Conceptual Understanding of Function

In this article, we have tried to explore the power and delicacy of a curriculum that encourages students to construct the concept of function as the foundation for learning algebra. We have tried to demonstrate a new role for the visual representation of function—as a model for building symbolic representation—with support from appropriate technologies that help bridge representations. In our work, we attempted to design a learning environment that would offer learners a coherent presentation of the concept of function—one focused on processes that encourage experimentation with mathematical objects related to representations of functions. To accomplish this goal, we chose rate of change as the primary characteristic of processes, and we chose the visual representation as the leading representation. We used a variety of technological tools to help students experience, abstract, and construct new variants of the representation of functions. Although the modeling of situations was the major activity, the environment was also designed to support experimentation with representations that do not necessarily emerge from modeling.

Analyzing the contribution of MBL technology to the construction of representations of function, we argued (as summarized in fig. 20.2) that although the links between a situation and its visual representation (and in a later stage, the link from symbols to situations) can be formed conceptually, other links (mainly from the graphs to the symbols) are quite fragile. Once students became proficient with visual representations, we hoped to design ways for them to use these representations as a basis for the construction of other representations. Figure 20.9 schematically describes the

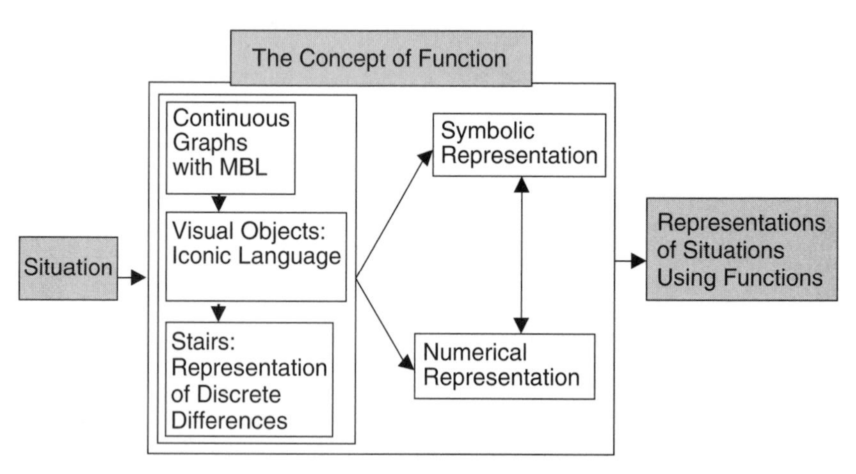

Fig. 20.9. Models of situations gradually linked to symbolic representations of functions

prealgebra curricular sequence whose major phases were described in this article.

When embedded in a prealgebra curriculum, the ideas presented in this article give a new dimension to problems used in context. They suggest ways to construct a symbolic representation of a function that models a situation by gradually linking the computer-made, continuous graph to the formal representation of the function. We believe that this type of learning then forms a strong foundation for algebra. We have found this approach very rewarding as we have struggled with the complexities involved in making the concept of function a central feature of the algebra curriculum.

REFERENCES

Arcavi, Abraham. "Symbol Sense: Informal Sense-Making in Formal Mathematics." *For the Learning of Mathematics* 14 (November 1994): 24–35.

Dennis, David, and Jere Confrey. "The Creation of Continuous Exponents: A Study of the Methods and Epistemology of John Wallis." *CBMS Issues in Mathematics Education* 6 (1996): 33–56.

Gravemeijer, Koeno, and Michiel Doorman. "Context Problems in Realistic Mathematics Education: A Calculus Course as an Example." *Educational Studies in Mathematics* 39, nos.1–3 (1999): 111–29.

Hazzan, Orit, and E. Paul Goldenberg. "Students' Understanding of the Notion of Function in Dynamic Geometry Environments." *International Journal of Computers for Mathematical Learning* 1 (1997): 263–91.

Heid, M. Kathleen. "A Technology-Intensive Functional Approach to the Emergence of Algebraic Thinking." In *Approaches to Algebra: Perspectives for Research and Teaching,* edited by Nadine Bednarz, Carolyn Kieran, and Lesley Lee, pp. 239–56. Dordrecht, Netherlands: Kluwer Academic Press, 1996.

Heid, M. Kathleen, with Jonathan Choate, Charlene Sheets, and Rose Mary Zbiek. *Algebra in a Technological World. Curriculum and Evaluation Standards for School Mathematics* Addenda Series, Grades 9–12. Reston, Va.: National Council of Teachers of Mathematics, 1995.

Hershkowitz, Rina, and Baruch Schwarz. "The Emergent Perspective in Rich Learning Environments: Some Roles of Tools and Activities in the Construction of Sociomathematical Norms." *Educational Studies in Mathematics* 39, nos.1–3 (1999): 149–66.

MacGregor, Mollie, and Kaye Stacey. "A Flying Start to Algebra." *Teaching Children Mathematics* 6 (October 1999): 78–85.

Nemirovsky, Ricardo. "On Ways of Symbolizing: The Case of Laura and Velocity Sign." *Journal of Mathematical Behavior* 13 (1994): 389–422.

Nesher, Pearla. "Microworlds in Mathematical Education: A Pedagogical Realism." In *Knowing, Learning, and Instruction: Essays in Honor of Robert Glaser,* edited by Lauren B. Resnick, pp. 197–217. Hillsdale, N.J.; Hove and London: Lawrence Erlbaum Associates, 1989.

Taylor, Peter D. *Calculus: The Analysis of Functions.* Toronto: Wall & Emerson, 1992.

Visual Mathematics: Computer Oriented Inquiry Curriculum for Junior High School and for High School. Tel-Aviv: CET, 1993, 1996. Also available on the World Wide Web at www.cet.ac.il/visual-math.

Yerushalmy, Michal, and Beba Shternberg. The Algebra Sketchbook. Computer software. Tel-Aviv: CET, 1993.

21

Teaching Mathematical Modeling and the Skills of Representation

Joshua Paul Abrams

IN A vital democracy, a primary goal of schooling should be the development of thoughtful, informed, and active citizens. Mathematics is an indispensable tool for reaching this goal. With mathematics, we can ask and answer important social, scientific, and political questions and analyze the claims that policymakers advance. This process of using mathematics to study questions from outside our discipline is called mathematical modeling. The methods and skills that we use to model a setting or situation provide us with a simplified representation, or model, of it that uses structures such as graphs, equations, or algorithms. The model, in turn, helps us generate new information about our original topic. During these interdisciplinary inquiries, mathematics is neither the motivation for our work nor the result. Instead, mathematics serves as an intellectual lens through which nonmathematical questions can be examined. When we teach our students to use mathematics in this way, we are providing them with an education that will serve them throughout their lives, to the benefit of society as a whole.

Many challenging and exciting skills used in the development of models of applied settings have been ignored by traditional school mathematics. These neglected skills include the abilities to analyze units, make choices among alternative forms of representation, and recognize common structures. These skills will receive proper attention only when we move away from a conception of school mathematics as self-contained and separate from other disciplines and move toward curricula that teach mathematics as a tool both for solving important problems from other disciplines and for making beautiful abstract discoveries.

A MODELING CURRICULUM

The habits and technical skills involved in creating representations of everyday phenomena must be explicitly taught. Students master them best in the context of an understanding of the modeling process. The process of modeling is best learned when students pose and explore their own questions about their world.

It is important to distinguish between open-ended modeling and application-oriented curricula. The latter use models and their contexts to motivate students to learn specific mathematics topics. They do not teach students how to model for themselves. By contrast, the open-ended modeling activities presented in this article help students practice the skills involved in creating mathematical representations. These activities are taken from a full-year modeling curriculum that has been taught in several high schools in Massachusetts. A portion of the curriculum, including handouts for students and software, is available at www.meaningfulmath.org. The course design and discussions are guided by the following essential questions:

- What is a mathematical model?
- What are some approaches to creating and modifying models of realistic situations? What are some of the different ways of representing a realistic situation? What are the benefits of each of these representations or mathematical tools?
- What knowledge can be derived from these models? How do we gauge the reliability of this information? In what ways can there be more than one right model or answer?

The course is carefully structured so that students identify questions that are important to them and learn ways of using mathematics to further their understanding of those questions. The culminating assessment of the course is based on a month-long, small-group project for which students pose their own questions and develop an original model. Sample questions that students have explored include the following:

- How does dirt affect the melting of snow mounds?
- What harvesting guidelines can protect a lobster population in the face of uncontrollable natural variations in the environment?
- How can delegates to the United Nations be seated to reduce tensions?
- How should a town structure its penalties for speeding tickets to generate the greatest revenue?
- What arrangement of ceiling lights provides the most even illumination for a room?

As these questions suggest, the goals of modeling endeavors vary. Modelers seek to gain understanding, predict outcomes, make decisions, and develop designs. The group examining the melting of snow mounds was inspired by the ever-dirtier mounds piling up during a snowy winter. The students simply wanted to understand what they were observing. They knew that the process of identifying variables, creating representations, working with the resulting abstractions to generate new information, and determining the significance of that information to the original question would help them toward that end. The group considering the plight of the lobsters worried that harvesting allowances might leave fewer than the minimum number of lobsters needed to replenish the stock. Group members developed a model for predicting changes in the lobster population over time. Using the model, they discovered that high harvesting limits would result in a complete collapse of the lobster population if a decreased birthrate or other perturbation reduced the stocks below the minimum level. The groups focusing on seating at the United Nations and speeding penalties both sought strategies that would optimize particular variables. The group examining illumination in the room was interested in a design that would improve the work environment of their windowless basement classroom.

THE MODELING CYCLE

Modeling is a cyclical process. Figure 21.1 details the steps, or stages, of the cycle, which many authors present in simplified schematics (see, for example, Clatworthy [1989]). The first stage, the creation of the mathematical representation, usually puts students on unfamiliar ground. Creating a model forces the modeler to think deeply about the setting. Translating an imprecise, complex, multivariate situation into a simpler, more clearly defined mathematical structure such as a function or a system of rules for a simulation yields several benefits. Modelers initially identify a list of variables. As they do so, they discover what they really know about their problem and what information they need to determine through library research or experimentation. In choosing a particular type of representation, they must think about the connections between and among variables, decide which relationships and structures are most important to capture mathematically, and pick the mathematical realm that offers the best possibilities for expressing all these features.

Just as a scale model, such as a model boat, represents a larger, more complicated object of the same type, capturing some of its aspects (appearance, proportions) while simplifying or neglecting others (size, materials, function), so too a mathematical model represents a situation by retaining aspects that are essential for study and putting aside details of lesser importance. Because of this simplification and because mathematical objects, such

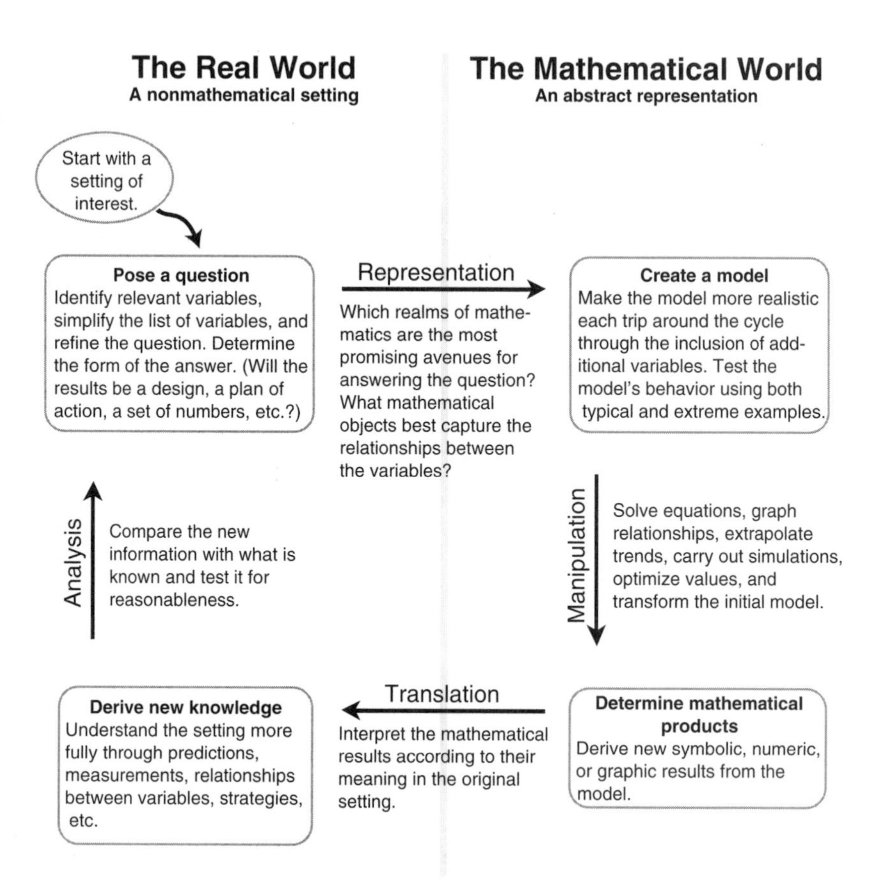

Fig. 21.1. The mathematical modeling cycle

as equations, have understood properties and behaviors, studying a mathematical model of a real-world situation can provide students with insights that are hidden from them during a nonmathematical study of the same situation.

Modelers must begin as simply as possible when they set out to create a manageable first model. For example, the group studying seating at the United Nations wanted delegates seated next to representatives from "friendly" nations. Rather than seat the delegates in a rectangular grid similar to that used by the U.N. General Assembly, the students seated their model delegates in a circle. This change simplified the setting through the introduction of symmetry. Every delegate now had exactly two neighboring delegates. The rectangular arrangement gave delegates seated in the interior of the grid four neighbors; those seated at the edge, three; and those seated at a corner, two. In addition, the change in shape led the group to consider their delegates as the vertices of a regular polygon. They then represented "enemy" relationships with line segments between the vertices and were able to apply their prior knowledge about polygons to the model. For example, because they wanted the sides of their polygon, which represented adjacent delegates, to be free of line segments representing enemy relationships, they could see that the number of such relationships among n delegates had to be no more than the number of diagonals in an n-gon or $n(n-3) \div 2$. Determining the number of permissible enemy relationships would have been much less straightforward for delegates arranged in a rectangular grid.

Once a first model is fully understood, additional variables can be incorporated into an increasingly complex rendering of it. This habit of initially excluding seemingly important aspects of a problem is counter to most students' instincts. The necessity of creating a preliminary representation, which does not fully solve a problem, needs to be illustrated, emphasized, and practiced repeatedly.

TEACHING THE MODELING CYCLE

The full-year modeling curriculum introduces the modeling cycle through an activity that requires only simple arithmetic yet allows students to practice the three main skills used in developing representations: identifying variables, simplifying, and choosing mathematical abstractions that capture concrete ideas. This activity provides a scaffolding for each of the individual modeling skills as they are learned throughout the course.

The activity involves a model that represents competing political parties (Hotelling 1929; Pollak 1980). The model demonstrates how different laws or voter behaviors may affect the changing political positions that parties adopt over time as they maneuver for advantage. The model is a simulation. A simulation requires the formulation of rules that describe the behaviors of the entities being studied. Once the simulation is carried out, the modeler can observe the consequences when those rules interact with one another.

The students are seated around tables in pairs. The teacher hands out a sheet with the simulation board and instructions (fig. 21.2) as well as a few small markers and a diagram of the modeling cycle shown in figure 21.1. After discussing the steps of the cycle, the teacher explains the political party model. The voters in this model are positioned along a spectrum from left to right (representing perspectives from liberal to conservative), and the parties also occupy positions along this political spectrum. Because we are in the United States, students start with two party markers. They place their markers down on two random spots. A marker at position 3 represents a party taking a moderately "left" position between the extremes. Students "run" the simulation according to the rules and see what happens over time. Figure 21.3 shows a sample simulation.

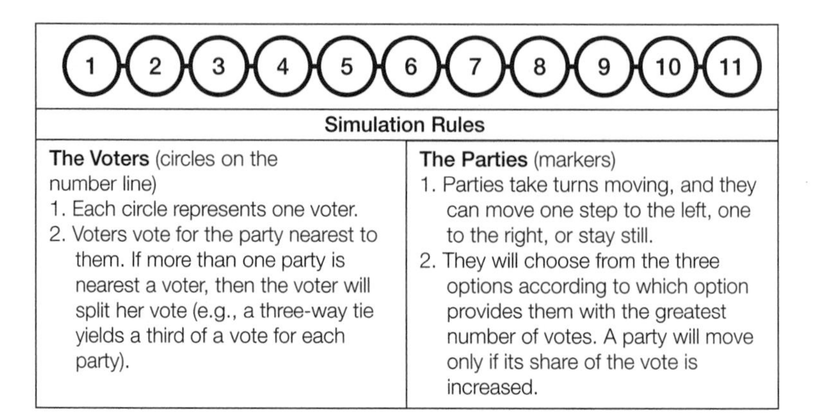

Fig. 21.2. The political parties handout

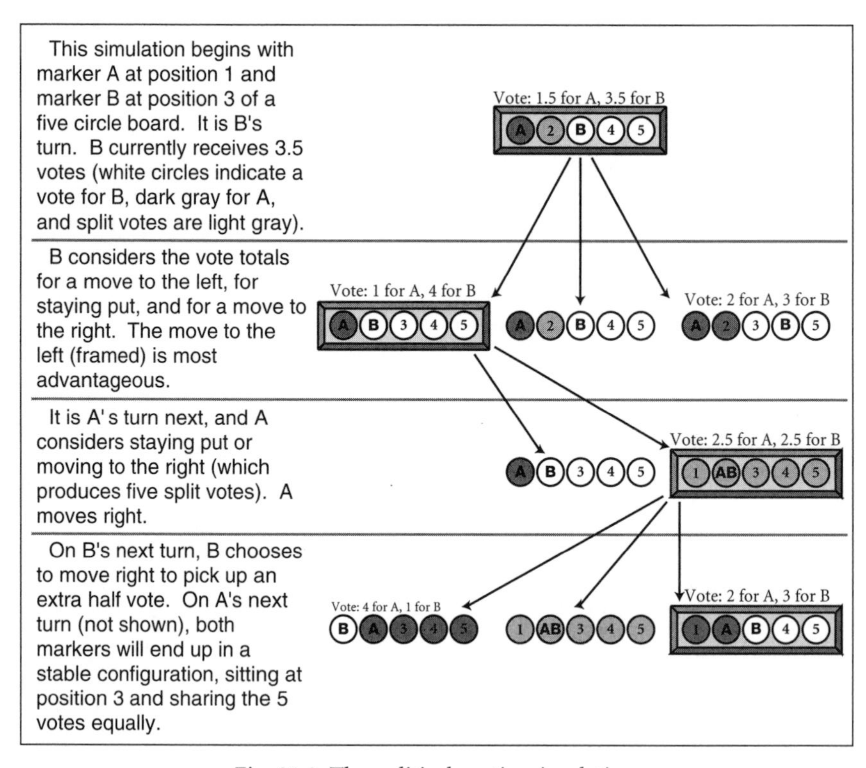

Fig. 21.3. The political parties simulation

TESTING REPRESENTATIONS

Soon after beginning the simulation, groups encounter the problems presented by adjacent and overlapping markers. For example, if marker A were at position 8 and marker B were at position 9, then A would receive 8 votes and B would receive 3 (from the voters at positions 9, 10, and 11, which were closer to it). If it were B's turn, then a move to position 8 would garner a total of 5 1/2 votes, since each marker would be equally close to all voters and all eleven votes would be split. A student's question, "Two markers can't be on the same number, can they?" immediately takes students into the thick of modeling decisions. First, they have to determine what an overlap of parties or its prohibition would mean in the real political setting. The teacher encourages students to choose whether they will allow such an overlap in the model or not, asking them to justify their decision on the basis of its fidelity to the real setting. Very different conclusions emerge from the two options. For example, if students prohibit the overlap because they believe that two parties would never have common platforms, then one marker can pin another one in an unfavorable position. The class gradually discovers that every aspect of their model should represent some aspect of the real-world situation.

Once the simulations have reached a stable configuration, the teacher seeks comments and observations. Students point out that when the parties are allowed to overlap, the markers freeze on the central circle. Students then interpret this mathematical result. At this point, they are engaging in the activity labeled "translation" in figure 21.1. They speculate that the positions of political parties will ultimately converge on a middle ground. When they compare this prediction from the model with what they know about political reality, they are participating in the activity identified as "analysis" in figure 21.1. The students engage in an animated exchange that reveals their varying beliefs about whether or not American political parties exhibit this convergence. Noting that the two parties in the model leave the simulated voters with no real choices, the teacher poses a follow-up question: "How many parties are needed to provide voters with distinct choices?" The class then explores the dynamics of multiparty systems by adding more markers to the simulation, and the students discover that at least four parties are required before the rush to centrist positions is reduced and voters have meaningful choices. The class compares these findings to the current impact of third parties in the United States. The students also discuss how the laws and constitutions of different countries influence the number of viable parties.

IDENTIFYING VARIABLES

During class discussions, students who contest the model's conclusions often begin to criticize its simplicity. Acknowledging that the lesson has presented a very simple model of a very rich phenomenon, the teacher asks the students to identify variables that the model originally ignored—that is, to examine differences between the model and reality. They see, for example, that real candidates' positions on many different issues give candidates complex political profiles. As students consider recent elections, they note that the model rests on many assumptions, including the following:

- Everyone votes, and voters are evenly distributed along the spectrum.
- People vote for a political position or a party platform and not for a specific candidate.
- Parties are always willing to change their views.
- A party's candidates and issues can be represented by one position on a single spectrum. A "pro-choice" fiscal conservative and a "pro-life" New Dealer would be difficult to position along the model's simple linear political map.
- There is no historical record, so having power does not secure future advantage.

Generating this list gives students practice in identifying variables—a skill that they need to initiate the modeling cycle. The list, which usually runs to twenty items or more, illustrates the necessity of starting with simple models. Students recognize the confusion that would result from trying to combine all their variables on a first attempt. Incremental improvements and subsequent analyses are more revealing and practical. To improve the model, students pick a variable that they would like to add to it and figure out how to incorporate this new variable into the current representation of the setting. For example, the student who offered the last item in the list of assumptions above suggested using a bigger marker for an incumbent party, identified as the one receiving the most votes on the previous turn, and awarding it a circle's entire vote in the instances that would previously have produced a split vote. This change in the model represented the advantage held by a party in power.

CREATING NEW REPRESENTATIONS

Representing missing variables effectively in the model requires students to practice looking for structures that both the nonmathematical setting and the mathematical entities have in common. For example, in the real world, incumbent parties win additional votes because of the benefits (greater visibility, patronage, control of redistricting) belonging to incumbency; analogously, in the model, a larger marker would now win extra votes according to the rule suggested by the student. The result of incumbency in both systems—real and modeled—was that weaker parties could not win by taking positions identical to those of their opponents. This process of developing an *isomorphism*, a mapping between two ideas with matching structures and parallel behaviors but often separate origins, is rarely experienced by high school students.

Modeling provides a context for teaching about isomorphism, which is an important feature of both applied and pure mathematical thinking. Students do not always attend to the most relevant features of a variable when creating representations. They need to ask themselves what characteristics are actually being captured by their proposals. For example, using a larger marker for incumbents proved helpful because the marker's size served as a reminder of its status, but the size itself did not affect the rules of the simulation. The marker could just as easily have been a different color, and the simulations would still have provided an isomorphic representation of incumbency. In contrast, a different student proposed representing an incumbent party with a dollar bill, which would cover two circles at a time, while coins would stand for the other parties. Although this proposal also contains a satisfying iconography, it is mathematically distinct from the first. The student's playful idea not only helps keep track of which party won the previous turn but also changes the geometry and the behavior of the simulation.

EVALUATING REPRESENTATIONS

Students invent mathematical representations for many of the new variables that the class identified. For each proposed change, they need to consider whether to alter the simulation's physical design or one of its rules, or both. They then try out the simulation with their changes and consider the meaning of the new behaviors that they observe.

Students' attempts to incorporate multiple political issues are particularly interesting. The students are quite intent on eliminating the assumption that "a party's candidates and issues can be represented by one position on a single spectrum." Some students suggest adding links between different voters, creating a network rather than a line. Others propose a grid of two or more dimensions in which each axis would represent an issue or a set of issues. Figure 21.4 Shows both proposals as well as questions that they raise. How do students choose between these alternative representations? They do so by paying attention to the consequences and coherence of each possibility. Both of these new configurations move beyond the geometric boundaries of a linear political spectrum. However, the first tangled network does not yield a clear mapping to the real situation. The new connections complicate the ordering of parties' positions, but students do not come up with an interpretation of this change that shows how it represents combinations of views on multiple political issues.

By contrast, the two-dimensional grid succeeds in representing every possible pairing of views, but students still need to make sense of how voters would make their choices (how distances between voters' and parties' posi-

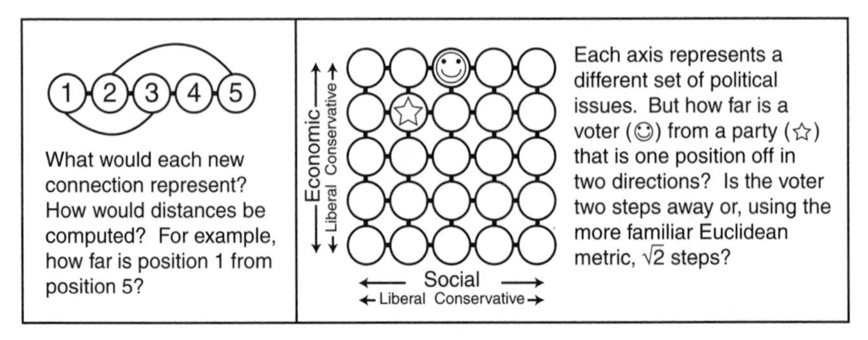

Fig. 21.4. Students' proposals for a nonlinear representation of political beliefs

tions would be measured). Should there be diagonals in the grid model? An informal discussion of metrics leads the students to compare beeline (Euclidean) and gridline (taxicab) distances. They wonder whether the metric is a crucial modeling choice influencing modeled party behavior. These considerations illustrate the rich questions about pure mathematics that can arise naturally from the complex issues and new directions suggested by modeling problems. Ultimately, students decide that the grid provides a consistent and improved representation for the model. Only through testing and by insisting that each part of a representation be interpreted (mapped onto a characteristic of the original problem) are they sure that they have made a good decision.

Students can also gauge the value of a model by looking at the predictions they can make from it. Do the mathematical results of the model translate into outcomes that are known to be true but that were not intentionally built into the representation? If so, then modelers should have an increased willingness to trust other newly discovered implications of the model that appear to have some degree of plausibility. For example, students who believe that real parties move toward the political center in the way that the simulated parties did are more comfortable accepting other interpretations drawn from the model.

LEARNING FROM THE MODELS

Students propose a variety of strategies for modifying the political parties model to represent less-than-universal voter participation. They suggest (1) randomly selecting voters to participate at each turn, (2) weighting the circles to reflect research on voter participation according to political beliefs, and (3) counting only those voters within a set distance from each party. (For example, voters at positions 10 and 11 might abstain from voting if the nearest party were at position 6.) By implementing each of these three proposals, students learn that some representations are more informative even when they may not appear to be the most realistic of the choices. The third suggestion has proved to be a particularly revealing option. If voters will only support parties within three units of their own position, then the markers that represent the two parties typically end up at positions 4 and 8. Thus, voters who consistently withhold votes from those parties that stray too far from a voter bloc can pull parties away from the center and assure distinct political choices. The mathematical model suggests that in a two-party system, voter behavior can achieve an advantage that having a greater number of parties offers in other democratic models.

In a few class periods, students are able to work through several cycles of critiquing and refining a model as they generate more realistic results. All students are able to create representations appropriate to their own mathematical sophistication. An exciting aspect of this unit is the students' surprise that mathematics can be used both to describe and to offer new understandings about political and philosophical issues such as voter behavior and the design of democracies.

REPRESENTING FUZZY CONCEPTS

Although modelers seek representations that clearly capture the meaning of a situation, they frequently confront intangible concepts that are challenges to express mathematically. Optimization is the most common goal in many students' modeling investigations. However, students often find it difficult to quantify the variable they want to optimize. For example, the student group studying speeding penalties wanted to maximize revenue, or the total amount of money raised, from tickets. That variable was easily described. In contrast, the group studying seating at the United Nations wanted to minimize tension. At the outset, it was not at all clear to the group how to capture the overall *tense*-ness attributable to a given seating arrangement. Tension proved to be a fuzzy concept that required a clear mathematical definition. Similarly, the *even*-ness of a room's lighting was challenging because providing even illumination called on the group to reduce variability over a continuous set—all the points in the plane of the floor. The students were familiar only with averages of finite lists of numbers.

The public frequently encounters mathematical measures of ill-defined concepts. *Consumer Reports* rates products' overall quality by combining many unrelated variables (such as the cost, quietness, and cooling ability of an air conditioner). *U.S. News and World Report* (Morse and Flanigan 2000) ranks colleges according to many quantifiable variables, some of which are of questionable merit. Recently, a mathematician was hired by New York City to create a function for measuring the social and mental well-being of the populace (Scott 1997). This type of "ranking" function is a necessary tool in modeling, and creating one provides a perfect means for students to practice many of the skills involved in representation and modeling. These include skills in posing problems, creating definitions, identifying variables, simplifying, making choices among representations, and justifying decisions.

Before each student creates his or her own ranking function as a long-term project, the class develops one together. One class considered several possible topics and ultimately decided to rank teachers. The variables that the students identified were a teacher's creativity, knowledge of the discipline, fairness, inclination to assign "busy work," experience, sense of humor, level of what they called the "monotony factor," hygiene, ability to give help, legibility of handwriting, number of court appearances, ability to challenge students, open-mindedness, opinionatedness, and sixty-two other factors! The class then tried to simplify their problem by grouping the variables by type (e.g., height, blood pressure, and gray hair were all combined in a physical variable) and rating them by importance. This second task forced them to clarify the matter of perspective—whose perspective should the ranking function take? Were they looking at the teacher from the point of view of a student sitting in the class or a principal hiring a new teacher?

The class spent a great deal of time evaluating the importance of a teacher's being to some degree opinionated (OP) or open-minded (OM). One student proposed the function $Rank_1(OM, OP) = OM \times OP$, with each variable being a subjectively determined whole number from 0 to 10. After investigating this representation of teacher quality, a classmate objected to the rule, noting that a very open-minded and not at all opinionated teacher would receive a score of: $Rank_1(10, 0) = 0$. This student suggested $Rank_2(OM, OP) = OM + OP$ instead. The first student replied that she preferred $Rank_1$ because she thought open-mindedness without any opinionatedness whatsoever would make for a weak and uninteresting discussion leader. She wanted high scores only when both characteristics were high. At this point, the students observed with surprise and genuine appreciation the correspondence between addition and the logical *or* and the analogous relationship between multiplication and the logical *and*.

During these conversations, the class came to recognize the different expectations and priorities that they brought to an evaluation of teachers. Some students sought an emotionally safe environment; others were more concerned with intellectual challenge. It was clear from the reactions of many that they had tacitly assumed that they held a common set of values. The process of mathematically defining a function forced them to articulate their differing views. As in the example of the geometric modifications of the political parties model, students arrived at an effective representation by generating multiple possibilities and studying the consequences of each in relation to their goals for the behavior of the model.

The class examined test cases (sample normal and extreme teachers) and studied the domain and range of the Rank_1 function. After one student pointed out that $Rank_1(4, 0)$ should not equal $Rank_1(9, 0)$, the class chose to modify the domain to exclude 0. Ultimately, the students decided that although both opinionatedness and open-mindedness were useful qualities for teachers to have to some degree, too much of either would become problematic. Figure 21.5 shows a student's sketch of the graph of the function that she wanted to use to filter both variables. She wanted both high and low scores to be penalized. The class then proposed different functions that would serve this purpose, and after narrowing their choices to two, they settled on $Filter(OP) = -|OP - 5.5| + 5.5$ rather than $Filter(OP) = -1/5\ OP(OP - 10)$. They chose to use absolute value instead of a parabola or sine wave because its linear behavior preserved the spacing of their original subjective ratings. This decision illustrates how context can prompt students to think deeply about the consequences of the properties of familiar functions. The final ranking function was then

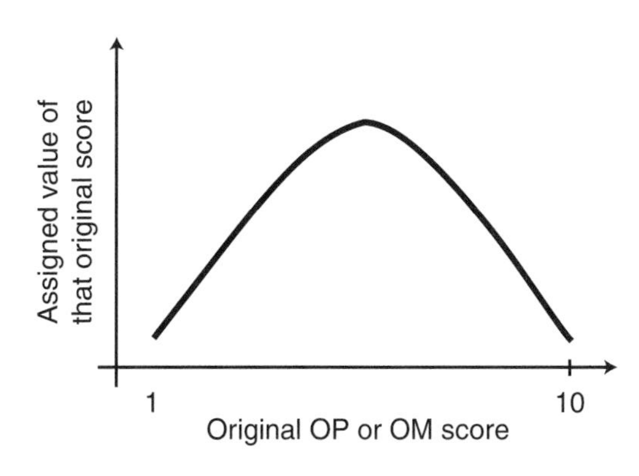

Fig. 21.5. A filter for teacher variables

$$Rank(OM, OP) = (-|OM - 5.5| + 5.5)(-|OP - 5.5| + 5.5)$$

with an output ranging from 1 to 25.

The process of building a ranking function parallels that of creating a whole model. Each new variable should be added iteratively, and each new stage should be tested fully. Each time the class was faced with a choice, they moved on only after a convincing reason had been given for choosing a particular representation. Of the original seventy-six variables, only a few were explicitly addressed by the final ranking function. They saw that after a certain point, the gain from adding new variables was offset by a loss in understanding that results from excessive complexity of the model. There is little doubt that a different group of students would have focused on a different set of variables. Thus, the modeling process combines attention to the internal logic and appropriateness of various mathematical structures with the subjective values and interests of the modeler.

In the preceding example, students' interest was quite high because of the freedom they had to choose a topic that was challenging, immediately relevant, and appealingly whimsical. The students used many basic mathematical skills involving the application and analysis of functions, but only as they themselves recognized the need for those skills. They were becoming increasingly self-directed as mathematical thinkers, and the teacher was able to assess how well each student was transferring his or her earlier learning.

CONCLUSION

How many of the habits identified in the lessons discussed here are mathematics skills? All of them are if as teachers we see our job as helping students make mathematics a vibrant part of their intellectual life. Representation is the first step in using mathematics to answer realistic questions. Teachers need to identify all the skills of representation and to provide their classes with a variety of contexts in which to apply them creatively.

REFERENCES

Clatworthy, N. J. "Assessment at the Upper Secondary Level." In *Applications and Modelling in Learning and Teaching Mathematics,* edited by W. Blum, J. S. Berry, R. Biehler, I. D. Huntley, G. Kaiser-Messmer, and L. Profke, pp. 60–65. Chichester, England: Ellis Horwood, 1989.

Hotelling, Harold. "Stability in Competition." *Economic Journal* 39 (March 1929): 41–57.

Morse, Robert J., and Samuel Flanigan. "How We Rank Colleges." *U.S. News and World Report.* 2000. www.usnews.com;usnews/edu/college/rankings/collmeth.html. World Wide Web (20 Dec. 2000).

Pollak, Henry O. "The Process of Applying Mathematics." In *A Sourcebook of Applications of School Mathematics,* prepared by a Joint Committee of the Mathematical Association of America and the National Council of Teachers of Mathematics (NCTM), pp. 1–9. Reston, Va.: NCTM, 1980.

Scott, Janny. "The Urban Happiness Quotient." *New York Times,* 10 June 1997, Sec. B, p. 1.